Military Innovation in Türkiye

This book explores Turkish military innovation since the Cold War. The major questions addressed are how Türkiye has been able to innovate, the production of new weapon systems, its philosophical background, how the country overcame bureaucratic and economic obstacles, and how these innovations resonated in military doctrine and organization.

Focusing on two main defense industry projects that trigger an overall change in the military doctrine and organization, the text examines the innovative inclinations of the Turkish military realm and reveals the societal, economic and political consequences of military innovation. This book fills a gap in the literature by providing an interdisciplinary and comprehensive overview of Turkish military innovation. Contributors include those involved in and affected by the military innovation process, as well as scholars who monitor the process using primary sources.

Military Innovation in Türkiye will appeal to academics, politicians and military professionals interested in understanding the evolution of the Turkish military.

Barış Ateş is an assistant professor at the Atatürk Institute of Strategic Studies (ATASAREN). He earned his PhD from Gazi University. His research focuses on military innovation, professional military education, and the sociology of the military profession. His research papers have appeared in international journals such as *Armed Forces and Society*, *İstanbul Journal of Sociology* and *Turkish Journal of War Studies*.

Routledge Military and Strategic Studies on the Middle East and North Africa

This series includes cutting edge research titles that study the use of force in the Middle East and North Africa. Topics to be covered include defence, terrorism, warfare, security studies, peace keeping, territorial claims and military innovation in these regions.

Military Innovation in Türkiye
An Overview of the Post-Cold War Era
Edited by Barış Ateş

For more information about this series, please visit: www.routledge.com/middleeaststudies/series/MSSMENA

Military Innovation in Türkiye
An Overview of the Post-Cold War Era

Edited by Barış Ateş

Routledge
Taylor & Francis Group

LONDON AND NEW YORK

First published 2023
by Routledge
4 Park Square, Milton Park, Abingdon, Oxon OX14 4RN

and by Routledge
605 Third Avenue, New York, NY 10158

Routledge is an imprint of the Taylor & Francis Group, an informa business

British Library Cataloguing-in-Publication Data
A catalogue record for this book is available from the British Library

ISBN: 978-1-032-35491-0 (hbk)
ISBN: 978-1-032-35493-4 (pbk)
ISBN: 978-1-003-32712-7 (ebk)

DOI: 10.4324/9781003327127

Typeset in Times New Roman
by SPi Technologies India Pvt Ltd (Straive)

To the innovators,
both civilian and military

Contents

Illustrations

Figures

Tables

Foreword

One of the most frequently quoted dictums that readers of military innovation come across is that generals often prepare for the last war and that is why armies pay a heavy price in the next war. Evidently there is a kernel of truth within this dictum, but most armies do nothing, and they certainly end up repeating most of the same mistakes.

Willingness to examine recent military performance with a critical eye is a very rare talent. Therefore, I must praise the editor and authors of this book for their courage in critically examining military innovation in Türkiye. The Turkish Armed Forces are famous for their culture of disciplined obedience, rigidity, and being slow to change. Additionally, the Turks do not like to discuss their military matters openly and frankly in public. Bringing Turkish military innovation to public debate, alone, is an important contribution.

The demand for military innovation in Türkiye has been stimulated by an uncertain future, frequent restrictions of arms sales, and even occasional embargos by the USA and European countries. Although access to funding and resources, rapid changes in domestic politics, and societal influences put serious barriers and limits on innovation, nevertheless, the Turks have achieved remarkable success in some areas. The chapters in this book not only provide essential information about innovation in Türkiye, but also explore the complexities and ambiguities.

Having edited collections of essays on military history and war studies and contributed chapters to other edited works, I am well aware of the challenges facing editors. Dr. Ateş does a terrific job as an editor by achieving a coherent structure and presenting fresh and original work to readers.

I cannot say that I agree with all the statements and findings of the authors, but they are in any case to be commended for the valuable insight and critical judgment they have brought to the subject. I believe the book as a whole will provide scholars and interested readers with a better understanding of military innovation in Türkiye.

April 17, 2022

Mesut Uyar, PhD, PD.
Professor of International Relations
Antalya Bilim University

Acknowledgments

I began researching military innovation during my doctorate in 2009. Reform, modernization, or restructuring were popular notions in Türkiye at the time; however, the same cannot be said for military innovation. In fact, I could not locate a single Turkish study that bore this title. Today's situation is not significantly different. Türkiye, which has considerably more intense military engagement and related experience than most other NATO countries, does not reflect the practice in academia. This book arose from a need during a period when the gap between practice and theory has been widening. First, the coup attempt, then counterterrorism operations, and finally, military operations in Syria and Libya received worldwide attention. This interest dramatically increased with the use of Turkish-made weapon systems during the Karabakh War. Finally, people witnessed Turkish-made armed drones in action in the Ukraine War during the publication phase of this book. However, this curiosity is not fully understood due to two overreactions: it is either used as black propaganda material against Türkiye or, on the contrary, embellished with a heroic narrative to praise the country and its armed forces.

It is fair to say that a few studies have looked into just the visible aspects of Turkish military innovation. However, these mainly mention armed drones and other weapons and equipment, but does Turkish military innovation solely consist of these aspects? How was innovation achieved? What were the results? To answer these questions, civilian and military experts held their first meeting two years ago at the Atatürk Institute of Strategic Studies (ATASAREN), Turkish National Defence University İstanbul. This is the first product of a modest step toward understanding Turkish military innovation, and we hope that it will inspire further research.

I would like to thank the contributors for their persistent dedication to completing the book on schedule. I also thank Mesut Uyar, who read this entire volume, contributed to its development with his advice, and showed kindness in writing a foreword. I also thank Beyzanur Arslan for her valuable contributions to improving the book.

I would also like to express my gratitude to Routledge's professional team, acquisitions editor James Whiting, and editorial assistants Euan Rice-Coates and Elizabeth Hart, for ensuring a smooth and instructive publishing process for me.

As a soldier-scholar who has been constantly away from home, nothing would have been possible without the love and faith of my wife and daughter, to whom I am eternally thankful. The views stated in this book are those of the individual authors.

May 13, 2022

Barış Ateş
ATASAREN

Notes on the Contributors

Barış Ateş is an assistant professor at the Atatürk Institute of Strategic Studies (ATASAREN). He earned his PhD from Gazi University. His research focuses on military innovation, professional military education, and the sociology of the military profession. His research papers have appeared in international journals such as *Armed Forces and Society*, the *İstanbul Journal of Sociology*, and the *Turkish Journal of War Studies*.

Mehmet Mert Çam is a research assistant at the Department of International Relations and Regional Studies, National Defence University. His research focuses on United States foreign policy and grand strategy studies. He graduated from St. George's Austrian High School. He earned his bachelor's degree in international relations at İstanbul Commercial University and his master's degree in war history and strategy from the Turkish War Academy. He is currently a PhD candidate in the Department of International Relations and Regional Studies, National Defence University.

Kemal Eker is head of the International Education and Cooperation Branch at Turkish National Defence University in İstanbul. He holds MA degrees in the science of strategy from Gebze Technical University and a PhD from Dokuz Eylül University, İzmir. His research includes naval warfare history, maritime strategy, history of communication technology, international migration, and political history.

Özgür Körpe is an assistant professor of military strategy at Army War College, İstanbul. He holds a BS from the Turkish Military Academy, MA degrees from the Strategic Research Institute and the Army War College, and a PhD from Yıldız Technical University. He has published books and articles on international security, strategy, and military planning on various platforms including *Military Review*.

Baybars Öğün is a research assistant at Ahi Evran University's International Relations Department in Türkiye. He earned his bachelor's degree from Yeditepe University, and master's and doctoral degrees from Gazi University. His research includes international relations (IR) theory,

Turkish foreign policy, and foreign policy analyses. He is also interested in political philosophy and political history.

Tolga Ökten is an instructor of Intelligence Studies at the Turkish National Defence University. His professional experience is on counterterrorism and counterintelligence. As a research fellow, he also worked at the Center for Strategic Research of the Ministry of Foreign Affairs, and at the NATO Centre of Excellence; Defence Against Terrorism. His research includes the conduct of war, intelligence, and terrorism.

Tolga Öz is an associate professor and the Vice-Rector at the University of Mediterranean Karpasia in the Turkish Republic of North Cyprus. His research involves the transformation of the Turkish defense industry, specifically after the Cold War era. He has published in journals such as the *Journal of Applied Security Research*, the *Journal of Defense Resources Management* (JoDRM), and the *Journal of Security Studies*. His research areas include: the defense industry, defense management, supply chain management, logistics management, aviation management, project management, and war studies.

Emrah Özdemir is an associate professor of international relations at National Defence University. He has a PhD in politics from the Department of Political and Cultural Studies at Swansea University, Wales. His research includes global politics, security studies, and war studies, specifically irregular warfare, and post-conflict reconstruction. He has published articles in journals such as *Interventions: Postcolonial Studies*, *Syrian Studies*, *E-IR*, *Independent Arabia*, and *LSE South Asia Centre Blog*, as well as chapters in books such as *Routledge Handbook of Peace, Security, and Development*, the *Encyclopedia of Violence, Peace, and Conflict (3rd ed.)*, and *The Future of the Middle East*.

Dağhan Yet is a research assistant at the ATASAREN Institute. He holds a BA from Koç University International Relations Department, MA degrees from both Boğaziçi University International Relations Department and National Defense University War Studies Department, and is currently continuing his PhD at the ATASAREN Institute. His research includes the philosophy of war, types of war, and strategy. In addition, he is an instructor in hybrid war at the War Studies Department.

Introduction
Turkish Versus Western Military Innovation

Barış Ateş

Introduction

Although military innovation studies began in the 1980s, real momentum started with the end of the Cold War. This was inevitable at this point because of the enormous pressure on armies to change, especially in Europe. During the Cold War era of a bipolar world order, where the threats and risks were obvious and predictable, states relied on mass armies supported by nuclear power. While offering nuclear security to their allies at the time, the United States and the Soviet Union demanded that they should maintain massive, mostly conscription-based armies to defend their national borders. After the collapse of the Soviet Union, conventional threats for continental Europe were replaced with smaller, unspecified risks (as opposed to obvious and predictable) that could originate from any part of the world. This led to the need for small but highly effective armies which could be mobilized quickly. In addition, societal factors, such as increasing levels of sensitivity toward casualties and conscientious objectors, also compelled armies to change. In this process, advanced technological weapon systems were a way to fill the emerging gap. Prioritizing peacekeeping operations over conventional warfare, these armies also had to professionalize.

However, it cannot be asserted that this process was carried out similarly in Türkiye. Various factors, including its geographical location and proximity to the endless conflicts of the Middle East, its different cultural and social characteristics, and most importantly it being the only NATO country fighting against a terrorist organization that threatens its territorial integrity have all caused military innovation to follow a different path in Türkiye. This is the book's precise objective: to explore and explain the Turkish military innovation process during the post-Cold War era. Of course, arguing that military innovation has only affected the military is not accurate. The effects of this change can be observed in various domains, from Türkiye's society to its international relations and economics.

How Türkiye has been able to produce new weapon systems, how those systems have affected its military organizational structure and doctrine, the philosophical background, how Türkiye has overcome bureaucratic and economic obstacles, and how these innovations resonate in other areas have not

DOI: 10.4324/9781003327127-1

yet garnered the desired attention. Meanwhile, studies from Western scholars on developed Western countries have gained momentum since the end of the Cold War in particular, and have examined these countries' military innovation in detail. In fact, not just political scientists and international relations experts, but also sociologists, economists, and historians have increased their interest in military innovation studies. Therefore, making Türkiye's military innovation movement the subject of a similar study has become a necessity for both academics, practitioners, and decision-makers.

Although Turkish military innovation has received more attention in recent years, this has been the result of projects and investments that date back much earlier. For instance, the production process of drones (UAVs) goes back at least two decades. However, new technology, as always, is expected to prove itself on the battlefield in order to be accepted by both soldiers and the rest of society. This opportunity has emerged recently in anti-terrorist operations within Türkiye and in the conflicts in Syria, Libya, and Nagorno-Karabakh. Since the Karabakh War in particular, this technology has begun to be discussed in international media and academic circles. In addition, the MİLGEM [National Warship] Project, which has gone through a production and development process as challenging as UAVs, as well as the missile systems and radars able to be used on this platform, have also been a focus of interest. Moreover, advanced electronic warfare systems; surface-to-surface, air-to-air, and air-to-surface missile systems; thermal imaging systems; tanks and artillery; and light weapons systems are also overwhelmingly indigenously developed.

The above explanations may allow one to conclude that military innovation in Türkiye has only involved technological advancements, and to some extent this might be an accurate inference. However, one should note that this book attempts to review Turkish military innovation from a broad perspective. The high profile of technological innovation is related to its results being clearly seen and attracting attention, both in Türkiye and the world. However, the academics' task is to provide a deeper understanding of the background factors. Therefore, this study aims to reveal the cause-effect relationship of military innovation in different spheres such as society, international relations, and the defense industry.

What Is Military Innovation?

Although military innovation efforts have been the subject of many academic studies, no consensus currently exists in the international literature regarding how to define these processes of change. In a quick review of the existing literature, various terms such as 'reform', 'change', 'transformation', 'modernization', 'restructuring', 'adaptation', and 'innovation' are used to describe the process. In parallel with this, the Turkish literature reflects the same ambiguity which does not just result in simple conceptual confusion. Each term can be perceived in different ways by individuals and institutions, and right from the start this leads to an incomplete or erroneous assessment of the processes of change.

To define the scope of the book, the terms 'transformation', 'innovation', and 'change' need to be clarified. Elinor Sloan, who has studied the transformation process of the US military, explains 'transformation' as a marked change in shape or quality, usually for the better. According to Sloan, today's concept of military transformation is actually the transformation of the concept of military technical revolution (MTR), which gained momentum in the 1980s, to the concept of the revolution in military affairs (RMA) in the 1990s, which then began being called military transformation at the beginning of the 21st century. One of the reasons for this change of terminology is that 'revolution' as a concept leads to high expectations. Therefore, the more modest term 'transformation' has been used.[1]

On the other hand, 'change' and 'innovation' are more inclusive concepts that are used in a more general framework. However, these terms can be used interchangeably. For example, Posen regards innovation as a significant change.[2] Zisk interprets doctrinal innovation as "a major shift in how military planners conceptualize and prepare for future war".[3] Rosen defines military innovation as "a significant change in the organization of armies for war and the way of fighting or the creation of a new combat branch".[4] In short, change is emphasized as needing to be comprehensive.

Another expert focusing on the magnitude of change is Grissom, who argues that in order for one to speak of true innovation, changes must directly impact combat units, and therefore, administrative or bureaucratic changes should not be considered as an innovation. Grissom's second criterion is related to the scope and impact of the change. Accordingly, small-scale reform movements with uncertain effects on the system should not be accepted as actual innovation. Finally, the change must provide a significant increase in military effectiveness. Thus, Grissom defines innovation as "a change in operational practices that provides a large increase in military effectiveness can be measured on the battlefield".[5] These definitions focus on the magnitude of change and reveal that Grissom, Posen, Zisk, and Rosen have not clearly distinguished between the terms 'innovation' and 'change' but instead use the concepts interchangeably.

Although Farrell and Terriff focus, like the others, on the difference between large and small change, they have defined 'change' and 'innovation' within clear boundaries. According to the authors, military change means "change in the objectives, actual strategies, and/or organizational structure of a military organization".[6] Small military change is less important because it has less resource impact on the military organization and state. Moreover, minor changes have less impact on international security as they do not involve adopting new military objectives, strategies, or structures. However, unlike the other authors, Farrell and Terriff accept and define the concept of innovation as one of three ways in which military change takes place, the other two being adaptation and emulation:

> Innovation involves developing new military technologies, tactics, strategies, and structures. Adaptation involves adjusting existing military

means and methods. Adaptation can, and often does, lead to innovation when multiple adjustments over time gradually lead to the evolution of new means and methods. Last, emulation involves importing new tools and ways of war through imitation of other military organizations. It is only when these new military means and methods result in new organizational goals, strategies, and structures that innovation, adaptation, and emulation lead to major military change.[7]

In short, innovation and major change are not synonymous for Farrell and Terriff. 'Innovation' is a method, and 'change' is a general term used to define the process and result. Accordingly, the main determinant is the addition of something *new* into the system. On the contrary, Sinterniklaas argues that innovation and adaptation are not easily distinguishable. He defines 'adaptation' as any sudden or gradual change in the way the military operates in response to the perceived or expected environment. 'Innovation', on the other hand, must have new concepts, methods, or new elements, such as weapon systems or other technologies. However, the concepts of adaptation and innovation are intertwined, and the distinction becomes controversial as no clear definition of *newness* exists.[8]

In light of the above discussions, this study also adopts the semantic meaning of 'innovation' that has been used extensively in recent years. Thus, 'innovation' can mean something that has never been discovered or implemented before as well as meaning something that makes changes to the existing structure.[9] In addition, every innovation more or less brings a change, but not every change can be said to involve innovation. For example, an organization may readopt a past process that had been shelved for some reason. This case also involves a change, but lacks innovation. Moreover, innovation must encompass critical military components such as doctrine, organization, culture, or equipment. As the above-mentioned experts emphasize, innovation must be large-scale. Therefore, I consider 'innovation' to be defined as movements that change all components of the military (i.e., doctrine, organization, technology, and culture) in order to acquire new capabilities or improve existing ones. The impact factor for each component might not be at the same level (e.g., organizational change might be major, but cultural change might be minor) depending on the nature and objective of innovation. However, their cumulative effect can be defined as 'innovation'.

In the case of Türkiye, many small-scale adaptation movements have spread gradually over time, as well as innovation movements that have emerged with sudden abrupt leaps. In general, small-scale movements are inconspicuous and only able to affect a small part of an entire organization. For example, personnel protection measures against mines in counterterrorism operations concern only ground troops, and even then just a small portion of these. In order to spread these innovations to the entire organization, additional conditions need to be created; however, the change here is gradual as these conditions differ in time and space. Meanwhile, the use of UAVs can change the entire military. Moreover, as this doctrinal change is based on battle-tested results, potential opponents within the organization have no opportunity to resist. However, if Türkiye had not experienced 40 years of terrorism threats or if no wars, such as Syria or Karabakh, had occurred, perhaps UAVs' current form would have reached nothing more than simple surveillance tools and thus not have been adopted at the same level. As a result, opportunities or challenging conditions need to be experienced before innovative methods are embraced.

Why Do Militaries Innovate?

Armies are claimed to be the most affected by national and international developments and must somehow keep up with them.[10] However, the first thing that comes to mind regarding armies is that they are large, traditional, closed-off structures that are resistant to change.[11] Of course, that militaries (whose main task is to fight, although they rarely do) are bound to their traditions and resistant to change should be considered normal. Analogies with other occupational groups may provide a better understanding of the subject. For example, throughout their careers, a surgeon carries out hundreds of operations, a teacher gives thousands of hours of lectures, and a cook prepares hundreds of different kinds of food. However, soldiers rarely participate in conflict or war, yet these are the main purpose for their existence throughout their careers; some armies experience no war or conflict at all. Therefore, soldiers and militaries do not have much reason to change.

Nevertheless, armies continue changing despite their lack of battlefield experience, and this tendency has drawn researchers' attention. As a result, military innovation research started in the 1980s, and the academic efforts that continued afterward have helped to reveal many of the factors that trigger innovation. Nonetheless, a generally accepted list has not yet been agreed on this subject. A recent study from Sinterniklaas sought to recapture technology, theater of operations, civil-military relations, international alliances and domestic politics, cultural values, and leadership as the key factors driving change.[12] However, this list does not mean the debate is over.

According to Barry R. Posen, the distribution of power within an organization also creates vested rights for employees. Therefore, individuals tend to oppose change within their organization as this will often be to the detriment of their vested rights. A similar problem exists in armies, as change may affect the vested rights of the officer corps. In addition to these problems, armies rarely utilize a change in military doctrine as this could increase operational uncertainty. Untested technology is also seldom expected to lead to a change in military doctrine, and may, instead, result in a revision to the old doctrine because technology that has yet to be tried and proven in a major war is often not allowed to change doctrine. As a result, armies enter the processes of change only after they or their allies suffer a major defeat, or as a result of pressure to adopt a new doctrine from civilians involved in military decision-making.[13]

Stephen Rosen has studied the innovation movements of the American and British armies and disagrees with the above arguments. As stated by Rosen, neither defeat in war nor civilian intervention can adequately explain the change. Interpreting military innovation as "an important change in how armies organize for war and battle," Rosen's study discusses innovation movements under three conditions: peacetime, wartime, and technological innovation. According to Rosen, the most important way to change an army in peacetime is through promotional policies. For innovation to occur, respected senior military leaders must believe significant structural changes exist in the international security environment before creating the new strategy accordingly. The next step requires increasing the promotional opportunities of the young officers who will keep up with the new strategy. Contrary to popular belief, the key to change is well-trained military personnel, time, and knowledge, not economic resources. Civilian intervention can be effective on this point as long as it protects the new military leaders.[14]

According to Kimberly Marten, who examined the innovation movements of the Cold War-era Soviet Army, soldiers generally want to use models that have been used and proven in the past, and therefore resist new ideas. However, exceptions may occur in which armies are able to carry out the change themselves under certain conditions. For example, if a planned change would enable an army to eliminate budgetary constraints and thus increase military autonomy, they would not oppose it. Therefore, armies can make doctrinal changes without civilian intervention. To explain this, Marten dwells on three separate arguments. First, soldiers often follow the doctrinal and technological developments of hostile armies and are always concerned about national security. This example explains how the US took and developed the concept of RMA from the Russians. She describes such innovation movements as 'reactive innovation': a significant change made by a potential enemy that makes the military question itself and ultimately innovate.[15]

The second argument is that, because not all soldiers think alike, they will have diverse reactions to innovation. The third argument is that direct civilian intervention in military change takes different forms and ultimately leads to bureaucratic conflicts at various levels. Therefore, an army should be persuaded rather than forced to innovate with an approach resembling 'crisis management'. Thus, the most significant change must first occur in the mindset of the personnel. As an additional remark, Marten highlights that innovative thinking often emerges in the military among those who do not have any corporate interests to protect, or who manage to put those types of career prospects in the background. For example, a study of new weapon systems in the US Navy concluded that innovators are middle-ranking individuals who are passionate, diligent, and in their early 30s to mid-40s.[16]

Another critical study that needs to be reviewed in order to adequately understand the subject is by Deborah Avant, who examined the experiences of the American and British armies in Vietnam, South Africa, and Malaysia. According to the author, civilian intervention is neither a sufficient nor necessary condition for change. The British Army was seen to have fulfilled the objectives set by the civilian authority despite the absence of civilian intervention, whereas the US Army was unable to achieve this despite civilian intervention. Connecting this situation to the flexibility and transparency arising from institutional structures, Avant tried to explain the roots of the problem through the approaches of the US and UK in the field of military history. In the past, the US Department of War was not very interested in developing military history and even refused to publish on issues related to US defense problems. On the contrary, British historians made significant contributions to the military education system. As a result, the Americans began to apply the scientific ideas developed by the Prussian Army directly without examining them in accordance with their priorities. At the same time, the British focused on how existing scientific knowledge could be used with respect to their military requirements. Therefore, the concept of safe leadership emerged in the US, in which leaders try to reach the summit unscathed by limiting their subordinates' initiatives and acting entirely on instructions for their career development. On the other hand, the British gave lower-level unit commanders the initiative, believing that every situation has its own characteristics that can only be perceived by unit commanders who have experienced the event in the field.[17]

Elizabeth Kier suggested that the relationship between change and military doctrine may best be understood from the cultural and political perspectives. Armies have different perspectives and perform their duties in different ways, and the institutional culture created by these elements directly affects which doctrine they will adopt. On the other hand, military doctrine is not usually designed to respond to threats from the international environment. The decisive factor on this point is the beliefs of the civilian decision-makers regarding the army's place in society. These beliefs affect a military's organizational structure. An internal political structure having no integrity during an ongoing conflict causes the military structure to be indistinctly determined. Armies often respond to the constraints resulting from such conflicts as dictated by their institutional culture.[18]

Kier explains the impact conflicting political-military subcultures have on the army using the experiences of France in the interwar period. At that time, the French Left wanted an army based on short-term conscription, while the French Right wished to establish a professional army. As a result of this debate, compulsory military service was reduced from three years to one year in 1928, which led to a sharp transformation in the army. In the early 1920s, the French Army had adopted an offensive doctrine and structure, and the entire army was shaped accordingly. However, the shortening of the military service period put pressure on the command level because the officers believed the one-year military conscription was only sufficient for a defense doctrine-based army. According to the officer corps, soldiers had to undergo years of training in order to maintain the current offensive doctrine. Therefore, they felt obliged to make a change in doctrine. Kier argues that the French Army had every means necessary to continue the offensive doctrine at that time, but the institutional culture did not allow it. In the same period, short-term military service was the most suitable structure for the German's offensive doctrine because they were able to form a large army by training hundreds of thousands of soldiers in a short time. Kier states that the main difference between the perspectives of two armies facing the same problem under the same conditions stems from their institutional cultures.[19]

Dima Adamsky also studied how cultural variables influence military innovation. He claims that the change depends on political, social, and technological factors, but none of these is sufficient on their own. Cultural factors constitute the pivotal intervening variable and become determinant in this process. In this sense, cultural factors may explain why some countries turn new technology into a true *military revolution* while others fail to do so or use new technology in very different ways. Adamsky argues that rationality, as neorealists suggest, is not always valid, and rational behavior depends on culture.[20]

After having reviewed the studies above, one is able to say that the question of why armies innovate cannot be adequately answered. Each expert puts forward a theory with different arguments and answers depending on the period and number of cases examined. Research sample size may be the main obstacle to presenting a generally accepted theory in this field. Moreover, explanations based only on developed Western countries (i.e., specifically using North American and Eurocentric examples) may remain less meaningful for the rest of the world. As a result, variables such as culture, historical factors, wars and conflicts, economic status, technological developments, civil-military relations, and perceived threats, which have been put forward as the factors driving military innovation, vary from country to country, even in deep-rooted defense alliances such as NATO. Türkiye might be one of the obvious exceptions within NATO as it has followed a different path with innovation. However, because the literature on Turkish military innovation is insufficient, this argument requires scrutiny.

Turkish Versus Western Military Innovation

In the post-Cold War era, the disappearance of the Soviet threat reduced the need for mass armies, especially in Europe, and new technological developments increased the demand for more trained professional soldiers. Moreover, the speed of war and the uncertain nature of threats led to the establishment of flexible and agile militaries. Another development that triggered changes in this period was globalization, as it decreased the importance of national borders so that no state could completely isolate itself from others. With the ever-increasing speed provided by the globalization phenomenon, moving information, money, goods, and services across borders became more straightforward. Consequently, interdependence between countries increased. In 1945, a single state had very few global problems that it could not solve by its own means. However, issues such as international terrorism, the proliferation of nuclear weapons, global pandemics, and lack of resources, today necessitate cooperation.[21] While this situation has reduced the likelihood of a great war between great powers, it has also led to the emergence of new risks such as environmental problems, hunger, and weapons of mass destruction.[22]

As a result, although the collapse of the Soviet Union in the post-Cold War period could be considered the most crucial reason for the changes in perceived threats, asymmetric threats originating from terrorism and migration as well as the increase in cybercrime, have also become important factors. The increasing economic interdependencies through the effect of globalization have additionally caused countries to restructure their armies in view of global considerations and regional factors. Military change became inevitable once these developments converged with technological advances.

The developments summarized above that took place after the Cold War led to significant debates on the role and duty of the army in Europe. The first and most striking result of these developments is that many Western armies reduced their number of personnel. The number of military personnel worldwide decreased from 28.8 million in 1987 to 22 million in 1997. As of 2021, this figure has decreased to 19.8 million. The most considerable decrease, 62%, was in Europe.[23] Despite the number of NATO member countries having increased, the number of NATO soldiers decreased by 40%, with this rate reaching 80% in some countries. In parallel with this development, 17 NATO countries either completely abolished or suspended compulsory military service. Furthermore, as no direct threat to any country's territory is expected for most European countries, protecting national borders is no longer identified as a primary task.[24] Instead, missions for international peacekeeping operations have begun coming to the fore.[25]

The same period has additionally brought new scientific developments, with high-tech weapons attracting the attention of many countries, especially after the swift victory and very few casualties in the 1991 Gulf War. The effects of technological developments on the character of war and armies are briefly defined as RMA.[26] According to Forster, European armies try to keep up with RMA by using battlefield information systems, sensors, and technologies that focus on precision strike capabilities. The use of civilian management techniques by the armies, combined with the civil and military technology, have additionally emerged as the focus of RMA.[27]

Meanwhile, the starting conditions and later developments in Türkiye have differed from those of other NATO countries. Despite being considered the second-largest NATO army in terms of number of military personnel, many examples can be found showing the Turkish Armed Forces (TAF) to have actually been in dire need of advanced technological systems. Prior to the end of the Cold War, Türkiye was defending 27% of NATO–Europe and 37% of the NATO–Warsaw Pact land borders.[28] In the same period, the Turkish Land Forces were comprised of 17 divisions and 23 brigades (a total of 520,000 personnel, 475,000 of whom were conscripts). The Turkish Navy had 55,000 personnel, of whom 42,000 were conscripts. The Turkish Navy only had 14 destroyers and four frigates. The Turkish Air Force had 55,000 soldiers, of whom 35,000 were conscripts. Also, the mandatory military service duration was 18 months.[29] However, almost all main weapons platforms were outdated, with some even having seen action in World War II. Therefore, the attempt was made to use manpower to balance out this deficiency.

Türkiye also differs from other NATO countries with regard to threat assessment and social indicators.[30] One of the main reasons for this is its proximity to conflict zones and the fight against terrorism, which has been going on for almost 40 years. PKK/KCK is a designated terrorist organization whose attacks caused the loss of more than 8,000 security officials between 1984 and 2020. The number of wounded in action over the same period is approximately 24,837. In addition, the number of civilians who lost their lives in terrorist acts was 5,700, with the number of injured at 11,347. Some operations in which a couple of generations of officers actively participated were at the scale of a typical conventional battle. Their experiences were crucial in initiating and, at the same time, embracing military innovation, as they had been able to witness operational problems in person. Social and cultural differences additionally manifest in military innovation movements in Türkiye in ways different than in the West. For instance, public support for the military has always been higher compared to most NATO countries. Moreover, the martial spirit that both soldiers and society possess does not follow other developed countries; for example, the Turkish military has never suffered a recruitment problem.

In summary, threat assessments, economic conditions, and even social and cultural factors have caused Türkiye to experience a different process of military innovation. As these examples have shown, Türkiye had to face different conditions in the post-Cold War period compared to other NATO countries. However, these differences have been consistently decreasing. Although conscription is still implemented, the current ratio of professional soldiers has reached almost 50%, which helps decision-makers reduce the total number of personnel by more than half. When looking at the main indicators, it is safe to argue that the manpower-oriented organization and doctrine having shifted toward a technology-oriented approach, thanks to both improvements in economic status as well as investments in military technology. Therefore, both similarities and differences between the Western and Turkish military innovation movements continue to affect the process.

The Content of the Book

Being multidimensional massive structures consisting of various elements such as personnel, materials, doctrine, culture, and expertise, armies are evidently unable to be explained through the efforts of any one discipline. In this regard, expertise in different fields such as political science, sociology, international relations, history, and economics is required. Therefore, the contributors to this book have been selected from among various disciplines. Additionally, some of them are soldier-scholars who have experienced Turkish military innovation firsthand. As one with considerable experience in military service, Özgür Körpe examines the philosophical background of the Turkish defense industry, arguing that Turks owe the practical and useful solutions they produce in times of crisis to their pragmatic cultural qualities, due to the geographical environment in which they live being in a state of constant crisis. For this reason, the efforts they make while overcoming crises result in benefits in terms of technological development.

Tolga Öz attempts to take a snapshot of the development of the Turkish defense industry and its capacity to support the TAF. He states Türkiye to have adopted a deterrence strategy based mainly on conventional weapons systems produced by NATO allies. However, in parallel with the rapidly changing environmental and technological global developments since the 2000s, this trend toward defense industry principles has been framed and updated to include domestic production, national self-sufficiency, and competence. This development was triggered by the sanctions, embargoes, and technical problems that can leave the country vulnerable to the dangers arising from separatist terrorism and other regional threats. Therefore, joint production, cooperation, technology transfer, and ending single-source dependency on supply/external procurement can be viewed as the main issues to be managed in order to increase autonomy within the next-generation defense industry strategy. To summarize, Türkiye seeks and strives to be less dependent on external sources, or, simply put, to be autarkic in the defense industry.

Chapters 3 and 4 focus on two important defense industry projects widely discussed in Türkiye and the rest of the world. Mehmet Mert Çam, a civilian scholar, attempts to explore the inside story of UAV development by analyzing it in light of the theory of bureaucratic politics. Eker, a soldier-scholar, scrutinizes the advent of the MİLGEM [National Warship] Project. The fact that one of the projects was initiated by civilians and the other by soldiers were essential factors in selecting these two projects. Despite this difference, both projects have common characteristics, and their mental development phases actually go much further back. Another important aspect is that these projects have acted as a catalyst by triggering innovation in other weapon systems.

Based on the conceptual and theoretical perspective of structural realism, Baybars Öğün focuses on the possible systemic effects of the uncrewed combat aerial vehicles (UCAVs) that Türkiye produces and uses effectively in its military operations. Öğün's main argument is that secondary states can create a *tendency to imitate* at the system level once they introduce military innovations within the scope of reinvention. Emulation is crucial for survival in international competition based on relative gains. This situation additionally has a relationship with socialization of the system. States that are able to transform emulations into an innovation stage can act more autonomously in their foreign policy, while encouraging other states to be at the same structural level. This is related to the fact that successful states at the system level create demonstrative effects. In addition, the emulation processes of secondary states are claimed to have become widespread after the Cold War by demonstrating how bipolarity at the system level limits internal balancing practices.

Emrah Özdemir examines irregular warfare as an innovative option, not just for non-state actors but also for conventional armies, by using different case studies, taking three historical examples in which irregular warfare has been innovatively and creatively used. The first example is the British experience behind the German lines during World War II. The second precedent concerns the practices falling outside the scope of the conventional warfare that the US Army tried to launch during the Vietnam War. The last example is the counterinsurgency operations from the coalition forces in Afghanistan and Iraq after the September 11, 2000 attacks. After examining these cases, the conceptual meaning of contemporary irregular warfare is explored. Finally, the TAF's changing approach to irregular warfare is examined in order to better understand its current significance. The main argument of this chapter is that irregular warfare has been evaluated as a novel and innovative option in times of stalemate, crisis, and despair. In a modern war environment dominated by hybrid and asymmetric conflicts, irregular capacities still present innovative opportunities as creative combinations of conventional and unconventional capabilities for conventional armies as well as non-state actors.

The impact of Turkish military innovation on the tactical level also requires analysis. For this purpose, Tolga Ökten focuses on two main components of military power: maneuverability and firepower. Innovations in firepower and maneuverability are observed to transform the doctrines of military organizations. The traces of this transformation process can be followed through changes in doctrines. For instance, the TAF has changed its doctrine from personnel-intensive search-and-destroy campaigns to an intelligence-driven find-fix-and-fight concept. This new concept achieves area denial not through quantity but through technology. Conventional engagements are conducted by a small number of agile volunteer forces supported by precision weaponry. Moreover, this transformation has not been limited to TAF. Other Turkish security agencies such as gendarmerie, police forces, and intelligence agencies have also transformed themselves.

Dağhan Yet focuses on the non-kinetic dimension of Turkish military innovation by examining approaches to information warfare (IW). The author argues every conflict requires a different approach toward IW in terms of pacing, for which he coins the term "Quick-Impact Approach (QIA)" to describe this concept. He chooses the recent examples of the Syrian conflict and the Second Karabakh War to illustrate the effect of QIA because, according to the author, the Turkish and Azerbaijani armies masterfully used QIA to support their kinetic victory on the battlefield. After examining and comparing these examples with failed ones, the author focuses on asking why QIA is vital for some conflicts and when the QIA approach should be used. The author believes Türkiye has been able to cope with terrorist organizations or third states' disinformation campaigns by effectively using drone footage. However, this kind of information capability relies solely on kinetic capabilities. Other tools that can provide accurate and timely information about Turkish military operations have not been sufficiently developed. Therefore, Yet concludes that Türkiye has not invested in non-kinetic means at the same level as kinetic means.

No book on military innovation would be complete without addressing the societal implications of military innovation. To fill this gap despite a lack of data on casualty aversion regarding Türkiye, I compare Western and Turkish societies' perceptions and support of their militaries and ongoing operations using the 'post-heroic warfare' theory. Social statistics and policies indicate that, although some signs are found indicating Turkish society to have entered a post-heroic phase, Türkiye still retains its heroic warrior beliefs. For example, the TAF has never had any recruitment problems, even during the years with high casualty rates. With the decreased birth rate and increased GDP and welfare compared to the 1990s, the profession of arms is still one of the most prestigious. Society's perceived threats appear to be one of the main factors differing Türkiye from other NATO countries.

To sum up, this book aims to reveal why and how Turks have carried out military innovation and how this innovation has been reflected in various fields. Of course, this study does not cover all relevant topics; however, it is expected to provide a basis for future studies and to trigger new research questions.

Notes

1 Elinor Sloan, *Military Transformation and Modern Warfare A Reference Handbook*, Praeger, Westport, 2008, pp. 7–8.
2 Barry R. Posen, *The Sources of Military Doctrine: France, Britain and Germany Between the World Wars*, Cornell University Press, New York, 1984, p. 47.
3 Kimberly M. Zisk, *Engaging the Enemy: Organization Theory and Soviet Military Innovation 1955–1991*, Princeton University Press, New Jersey, 1993, p. 4.
4 Stephen Peter Rosen, *Winning the Next War: Innovation and the Modern Military*, Cornell University Press, Ithaca, 1991, p. 7.
5 Adam Grissom, "The Future of Military Innovation Studies", *Journal of Strategic Studies*, 29:5, 2006, p. 907.
6 Theo Farrell & Terry Terriff, "The Sources of Military Change", in (eds.) Theo Farrell and Terry Terriff, *The Sources of Military Change Culture, Politics, Technology*, Lynne Rienner Publishers, London, 2002, p. 5–6.
7 Farrell & Terriff, "The Sources of Military Change", p. 6.
8 Rob Sinterniklaas, "Military Innovation: Cutting the Gordian Knot", *Research Paper:116*, Netherlands Defence Academy, Breda, 2018, p. 30.
9 *Webster's Encyclopedic Unabridged Dictionary of the English Language*, Gramercy Books, New York, 1996, innovation, p. 984; change, p. 344.
10 John A. Williams, "The Postmodern Military Reconsidered", in (eds.) Charles Moskos, John A. Williams, David R. Segal, *The Postmodern Military: Armed Forces after the Cold War*, Oxford University Press, 2000, p. 265.
11 Rosen, *Winning the Next War*, p. 2.
12 Sinterniklaas, "Military Innovation: Cutting the Gordian Knot", p. 31.
13 Posen, *The Sources of Military Doctrine*, p. 57–80.
14 Rosen, *Winning the Next War*, p. 5–21.
15 Marten-Zisk, *Engaging the Enemy*, p. 3–10.

16   Marten-Zisk, *Engaging the Enemy*, p. 11–25.
17   Deborah D. Avant, *Political Institutions and Military Change: Lessons from Peripheral Wars*, Cornell University Press, New York, 1994, pp. 17–27.
18   Elizabeth Kier, "Culture and Military Doctrine: France Between the Wars", *International Security*, 19:4, 1995, p. 66.
19   Kier, "Culture and Military Doctrine", pp. 72–77.
20   Dima Adamsky, *The Culture of Military Innovation: The Impact of Cultural Factors on The Revolution in Military Affairs in Russia, The US, and Israel*, Stanford University Press, California, 2010, pp. 10–11.
21   Shannon L. Blanton & Charles W. Kegley, *World Politics: Trend and Transformation*, 2016–2017 Edition, Cengage Learning, Boston, 2017, p. 294.
22   J. Baylis, "International and Global Security in the Post-Cold War Era", in (eds.) John Baylis and S. Smith, *The Globalization of World Politics: An Introduction to International Relations*, Oxford University Press, 2001, p. 269.
23   Ljubica Jelusic, "Conversion of the Military: Resource-Reuse Perspective after the End of the Cold War", in (ed.) Giuseppe Caforio, *Handbook of the Sociology of the Military*, Springer, New York, 2006, p. 353. "Country comparisons", *The Military Balance*, 121:1, Routledge, London, 2021, p. 522.
24   Anne Aldis, "Defence Transformation in Europe Today: Implications for the Armed Forces", eds. Timothy Edmunds and Marjan Malešič, *Defence Transformation in Europe: Evolving Military Roles*, IOS Press, Amsterdam, 2005, p. 104.
25   Timothy Edmunds, "A New Security Environment? The Evolution of Military Roles in Post-Cold War Europe", eds. Timothy Edmunds and Marjan Malešič, *Defence Transformation in Europe: Evolving Military Roles*, IOS Press, Amsterdam, 2005, p. 10.
26   Andrew Latham, "Warfare Transformed: A Braudelian Perspective on the 'Revolution in Military Affairs", *European Journal of International Relations*, Vol. 8:2, 2002, p. 233.
27   Anthony Forster, *Armed Forces and Society in Europe*, Palgrave Macmillan, New York, 2006, p. 5.
28   "Defence and Economics in Turkey", *NATO's Sixteen Nations*, Special Issue, Vol 31, Brussels, September 1986, p. 40.
29   "Defence and Economics in Turkey", pp. 85–98.
30   For a general overview see: Bill Park, "Defence Transformation and Internal Security: The Turkish Experience", in (eds.) Timothy Edmunds and Marjan Malešič, *Defence Transformation in Europe: Evolving Military Roles*, IOS Press, Amsterdam, 2005, pp. 91–101.

Bibliography

Adamsky, Dima, *The Culture of Military Innovation: The Impact of Cultural Factors on The Revolution in Military Affairs in Russia, The US, and Israel*, Stanford University Press, California, 2010.
Aldis, Anne, "Defence Transformation in Europe Today: Implications for the Armed Forces", in (eds.) Timothy Edmunds and Marjan Malešič, *Defence Transformation in Europe: Evolving Military Roles*, IOS Press, Amsterdam, 2005, pp. 103–111.
Avant, Deborah D., *Political Institutions and Military Change: Lessons from Peripheral Wars*, Cornell University Press, New York, 1994.
Blanton, Shannon L. and Kegley, Charles W., *World Politics: Trend and Transformation*, 2016–2017 Edition, Cengage Learning, Boston, 2017.
"Country comparisons", *The Military Balance*, 121:1, Routledge, London, 2021.

"Defence and Economics in Turkey", NATO's Sixteen Nations, Special Issue, Vol 31, Brussels, September 1986.

Edmunds, Timothy, "A New Security Environment? The Evolution of Military Roles in Post-Cold War Europe", in (eds.) Timothy Edmunds and Marjan Malešič, *Defence Transformation in Europe: Evolving Military Roles*, IOS Press, Amsterdam, 2005, pp. 9–18.

Farrell, Theo and Terriff, Terry, "The Sources of Military Change", in (eds.) Theo Farrell and Terry Terriff, *The Sources of Military Change Culture, Politics, Technology*, Lynne Rienner Publishers, London, 2002. pp. 3–20.

Forster, Anthony, *Armed Forces and Society in Europe*, Palgrave Macmillan, New York, 2006.

Grissom, Adam, "The Future of Military Innovation Studies", *Journal of Strategic Studies*, 29:5, 2006, pp. 905–934.

Jelusic, Ljubica, "Conversion of the Military: Resource-Reuse Perspective after the End of the Cold War", in (ed.) Giuseppe Caforio, *Handbook of the Sociology of the Military*, Springer, New York, 2006, pp. 345–360.

Kier, Elizabeth, "Culture and Military Doctrine: France Between the Wars", *International Security*, 19:4, 1995, pp. 65–93.

Latham, Andrew, "Warfare Transformed: A Braudelian Perspective on the 'Revolution in Military Affairs", *European Journal of International Relations*, 8:2, 2002, pp. 231–266.

Marten-Zisk, Kimberly, *Engaging the Enemy: Organization Theory and Soviet Military Innovation 1955–1991*, Princeton University Press, New Jersey, 1993.

Park, Bill, "Defence Transformation and Internal Security: The Turkish Experience", in (eds.) Timothy Edmunds and Marjan Malešič, *Defence Transformation in Europe: Evolving Military Roles*, IOS Press, Amsterdam, 2005, pp. 91–101.

Posen, Barry R., *The Sources of Military Doctrine: France, Britain and Germany Between the World Wars*, Cornell University Press, New York, 1984.

Rosen, Stephen P., *Winning the Next War: Innovation and the Modern Military*, Cornell University Press, Ithaca, 1991.

Sinterniklaas, Rob, *Military Innovation: Cutting the Gordian Knot*, Research Paper:116, Publications Faculty of Military Sciences Netherlands Defence Academy, Breda, 2018.

Sloan, Elinor, *Military Transformation and Modern Warfare A Reference Handbook*, Praeger, Westport, 2008.

Webster's Encyclopedic Unabridged Dictionary of the English Language, Gramercy Books, New York, 1996.

Williams, John Allen, "The Postmodern Military Reconsidered", in (eds.) Charles Moskos, John Allen Williams, and David R. Segal, *The Postmodern Military: Armed Forces after the Cold War*, Oxford University Press, Oxford, 2000, pp. 265–277.

1 Turks' Pragmatic Solutions

The Philosophy Behind Defense Technology

Özgür Körpe

Introduction: The Chicken and the Egg – Need Versus Technology as a Causative Issue

Do operational needs develop technology, or do technological developments guide operational methods? Developments in the Turkish defense industry in recent years have brought these and similar chicken or the egg problems to the agenda. A symbiotic relationship exists between technology and operational methods. In other words, just as operational levels express the needs faced in the field, so, too, does technology, as driven by the strategic level, develop systems for responding to these needs. However, technologies that often go beyond the expectations of those in the field can also have effects that lead to a change in operational methods up to a point. In short, the operational part of this symbiosis could be called the trigger, while the technology part could be called the producer.

Today's battlefield has taken on a complex, fuzzy character which often produces multidimensional crises. Humankind is being swept from one deadly crisis to another. In such an environment, the explanations of modern political philosophy are insufficient, and destructive and deadly problems are now encompassed within the borders of philosophy. At this point, pragmatism, as a limitation problem of political philosophy, has the capacity to produce valuable solutions for fuzzy and wicked combat problems. Throughout the entire industrial age, strategic-level problems have essentially always been wicked. Therefore, the strategic level has already been accustomed to multidimensional, complex, and fuzzy decision-making methods. Once all environmental factors are excluded, the decision-making processes of an early 20th century strategic political/military decision-maker and their early 21st century counterpart generally have pragmatic similarities. What makes today's problems different is that this complexity has expanded to the operational and tactical levels. Now, the decisions made at the tactical level have effects at the strategic level, and the strategic level often interferes with the tactical level. Therefore, nothing strange is found in the concepts *strategic corporal* and *tactical general* becoming simultaneously popular.

Defense technologies are the primary environment in which this intricate situation is set. Now a minister of defense or a senior commander can view

DOI: 10.4324/9781003327127-2

real-time footage of a tactical-level conflict thousands of kilometers away, command and manage the conflict, and guide the firing of precision engagement platforms such as air force units, armed uncrewed aerial systems (A-UASs), and even artillery. On the other hand, a bullet hitting a nearby civilian settlement during a conflict between a commando team and a few terrorists may be reported in the international media for days. In summary, today's conflict environment has a hybrid character that produces multidimensional, fuzzy, and wicked problems at the strategic, operational, and tactical levels. The pragmatism here attributed to Turkish strategic culture fills the philosophical gap in solving these wicked problems. Many authors have pointed to the pragmatism in Turkish strategic culture, albeit in different contexts. The most important thing is to be able to explain creative decision behaviors at operational and tactical levels, within the framework of pragmatism.

Hoffman attributed the pragmatic nature of hybrid warfare to Lawrence and referenced his claims that the cognitive field would gain more importance than at any time in history in the future battlefield.[1] This environment dictates practical decisions as it is one where conditions change rapidly; friends, enemies, and neutrals become more uncertain and changeable; and sophisticated methods are increasingly preferred.

In this context, the central argument of this chapter is that the utility function of technology connects it to Türkiye's defense technologies through pragmatism, because development in Türkiye's defense technologies in recent years has been fed by the pragmatic character of Turkish strategic culture. It should also be emphasized from the outset that this chapter is not intended to be a eulogy to pragmatism. Instead, this chapter attempts to give the most appropriate name to the causality behind a strategic behavior that has attracted attention and aroused curiosity in recent years. In this context, the chapter will first discuss pragmatism's ability to solve problems and then reveal the causal relationship between technology and field needs. Without getting bogged down in details, the chapter will next question how much technology is able to affect the Turkish Armed Forces' fighting style, and finish by emphasizing the risks of technological development, which will also provide ideas for further research. Considering the anecdotes that are included here from the literature review as participant observations may help readers understand the logical flow of the chapter.

Pragmatism as an Explanatory Instrument

Technology is a phenomenon with which pragmatists have sympathized from the very beginning. Among the classical pragmatists, John Dewey is the one whose writings most closely resemble the concerns of the technologist. Metaphysical thought has little value in Dewey's pragmatism, which sees the human mind as having an instrumental position in finding effective solutions to the problems people face in their daily lives, both today and in the future.

According to Dewey, this is because confusing questions about sources need to be avoided.

For this reason, Dewey's understanding of knowledge brings with it an understanding of natural science and nature different from that of classical philosophy. According to Dewey, no science is prioritized, including physics.[2] Instead, the different ideas, concepts, and methods used in various branches of science are just tools humans use to reach their own goals and solve the problems they face.[3] Moreover, these tools vary according to the problems humans are trying to solve. This epistemological approach of Dewey's is called *instrumentalism*.

The instrumentalism Dewey developed shows his affinity to the philosophy of technology. For Dewey, not only are physical tools instruments, but so are concepts and methods; therefore, his understanding of technology is comprehensive. In fact, according to Dewey, there is no opposition to be found between technology and the rest of culture. Dewey wished to consider all traditional philosophical dichotomies, facts, and values, mind and body, and thought and action not as absolute distinctions of genre but as poles in the continuity or stages of activity.[4] This aspect of Dewey's philosophy fits the need for technology to deal with the ethics, concepts, and theory of action as well as science.

Dewey is important in terms of the pragmatic nature of Türkiye's defense technologies because traces of Dewey are observed in the founding philosophy of the Republic of Türkiye. Dewey came and stayed in Türkiye for two months in 1924. During this time, he met with Atatürk once, examined the education system, and wrote two reports.[5] During this period, he came into contact with many intellectuals, bureaucrats, and academicians.[6] In Dewey's reports, a realistic, life-capturing, and solution-producing approach clearly manifested itself as one where students should be connected to the life around them, where middle school graduates should be prepared for life and choose a profession even if they could not enter college, where children should not be burdened with their memories alone, and where practical skills should be given importance.[7]

For these and similar reasons, Kırık and Morva explained the philosophical background of the foundation of the Republic within the framework of *pragmatic action theory*[8] based on Atatürk's ideas and actions. Here, they also claimed that pragmatism has the capacity to explain the effect causality has on the tendencies present in Turkish strategic culture. In this context, pragmatic strategic culture has lain behind the rapid development of defense technologies in recent years.

Pragmatic thinking considers that theories require a counterpart in real life, to be applicable, and to have an effect once implemented. This is a unique perspective in terms of defense technologies. War is a fact of life. Warriors are in direct dialogue with the actual and painful reality throughout the campaign. For this reason, they try to reduce the pain to which they are exposed by defeating the enemy with the least casualties. This is what Sun Tzu said millennia ago: "Hence to fight and conquer in all your battles is not supreme excellence; supreme excellence consists in breaking the enemy's resistance without fighting".[9]

According to pragmatic thinking, war should also be completed as soon as possible by using the deadliest blows on the enemy. Combined force structures and combined effects should be created in order to do this. The commanders of mixed forces put the enemy in an inextricable situation by combining the different capabilities of various forces in a complementary and supportive way. Thus, as the enemy flees from the effects from one activity or system, they get exposed to another. This situation paralyzes the enemy, destroys them, and forces them to surrender.

On the other hand, the way that today's strategic problems have become more complex compared to the past, and constantly produce crises should not be overlooked. Hence, pragmatism is more of a necessity than an advantage because pragmatic solutions, by their very nature, usually come into play in times of crisis requiring a quick but useful response. The reason why war, a boundary problem of political philosophy, is in close contact with pragmatism is that war itself is a crisis.

Troop-Made Solutions at the Edges of Political Philosophy

The early years when I started my duty as an officer coincided with the period when Türkiye was carrying out large-scale, cross-border operations against the PKK terrorist organization based in northern Iraq, who had taken advantage of the authority vacuum after the First Gulf War. In those years, the Turkish defense industry had limited production under the umbrella of NATO. As a result, Turkish defense industry products were mainly imported from all countries that allowed them to be purchased through their patents. One of these was the 60 mm commando mortar, a domestic product.

The USA-made 60 mm M2 mortar, which had been part of the inventory since the 1950s, was heavy due to its simple bipod, and also caused lost time in conflicts with terrorists in mountainous regions as well as accidents due to the rocky and sloping ground. Moreover, this mortar required a standard mortar crew, which meant four people were engaged with one tiny gun. To increase the functionality of M2 mortars, its bipod was removed during counterterrorism operations, with only its barrel being used. However, because the collimator provides elevation control on the bipod, practical solutions such as triangular wedges and knotted ropes were developed to give proper elevations. Of course, these solutions did not provide sufficient benefits. For this reason, new 60 mm mortars were imported from Spain. However, it took more time for the conscripted soldiers to understand the instructions for the collimator and elevation controls of the Spanish mortar compared to the previous one.

After all these negative experiences, the Mechanical and Chemical Industry Corporation (MKEK) began producing a domestic commando mortar. The domestic mortar removed the bipod from the American M2 mortar, as well as its mounted collimator and sightline, thus eliminating the need for the crew, who had been mostly idle before. The disc-shaped elevation of the Spanish mortar was packed into the carrying handle with a smaller, more practical mechanism. However, the sightline, which had been inspired by the Spanish mortar, was problematic in terms of accuracy. Although the mortar was designed to be used by two people, the gunner had to sit down and hold the mortar so that it would not deviate from the target after aiming, while the other personnel loaded it. But line of sight would still get broken by inevitable human reflexes. Problems increased when the flanks were hit: the commanders began allocating a third person to control the line of sight from behind. This third person, two or three steps behind the gun, would line up the gun and the target, aim with a compass, and check the accuracy of the aim line. When shooting prone, a third person became even more necessary. This was not a practical solution. Therefore, I and my teammates manufactured a simple sight from a firing pin and an empty mortar bomb-tube and started using the weapon this way. The hit rate immediately increased by 80%.

Years later, I came across different versions of this simple mortar sight among other troops. Of course, that our previously developed model had reached these units was unlikely, and similar sights had probably been produced using similar logic. However, the striking aspect of this process was the plainness of the developed solution. This, and many other similar practical solutions can still be found in the Turkish Army today, as well as many officers and non-commissioned officers who claim to have discovered these. Moreover, this type of "freehand"[10] equipment is not commonly found in any other Western army. Soldiers producing something with scarce outmoded tools from the barracks is certainly nonsensical, ineffective, and unsustainable, yet troop-made solutions are undoubtedly pragmatic because they serve their purpose by bridging the gap between slow-moving logistics responses and vitally urgent and rapidly developing operational needs. Thus, pragmatism shows itself once again at the edges of political philosophy.

Pragmatic Solutions to Wicked Problems: Lathing Full-Round Projectiles

Defense technologies are effective at solving fuzzy and wicked problems. In fact, having technological superiority provides a strategic advantage beyond simply problem-solving. Throughout history, the states that develop new technologies are the ones that succeed at establishing dominance over other states or deterring threats against themselves. For example, the Hittites produced jugs by shaping clay and thus were able meet the water needs of their armies and expand their operational capacity to Mesopotamia and Palestine. Likewise, thanks to Archimedes' lethal war machines, the Syracusans were able to shatter the Roman ships that came to invade their island.

Meanwhile, the Romans gained significant advantage over their enemies thanks to their huge ballistae. By improving cannon technology, the Turks were able to tear down and pass the wide and high walls of İstanbul.

In addition to their superior artillery, Turks had improved volley fire by developing the tactical use of the musket, thus achieving victory in Mohacs in a few hours, conquering Cyprus with the effective use of the arquebus, and finally gaining undisputed superiority across a large part of continental Europe from the mid-15th to the mid-17th century. When the British Empire added steam power to its warships, it consolidated its naval supremacy for another century. By developing the atomic bomb, the United States of America gained an undisputed strategic advantage globally. Countless examples can be given. However, what is emphasized here is that developing defense technologies only at the tactical level is not enough. A level needs to be reached by producing technologies that will provide a strategic advantage unavailable in other states.

Türkiye's developments in the field of defense technologies in recent years have also attracted global attention and appreciation. So, what makes them pragmatic? The answer to this question is directly related to the purpose of the military operation. When the aim is to terminate the operation in order to realize national interests, defense systems that are imported and borrowed may be a vulnerability. In this case, the most pragmatic solution would be to develop one's own technology. To avoid the time and labor costs of producing this technology, a more pragmatic solution may appear at first to buy the produced technology instead of producing it.

However, buying technology directly from outside inevitably results in bearing the usage restrictions of the producing country; paying a high price for the product due to the conditions of a monopolistic competition market or even oligopoly market; being dependent on the producing country to supply the spare parts for maintaining and repairing the product; and being constrained to a technological production with regard to the language and culture of the inventing country. To adapt to the product, one must inevitably be willing to change one's behavior (i.e., be trained by that country).

All these possibilities may discourage one from importing from the very beginning. Therefore, importing products from abroad may be asserted as being unable to serve the purpose of an operation based on realizing national interests. In short, a close relationship exists between pragmatic purposefulness and national technology. In fact, Turks throughout history have experienced being deprived of developing the strategic benefits of technology and having to endure the strategic losses of being dependent on foreign sources. Illustrating by giving an example from history would be useful at this point.[11]

During the critical days of the Turkish War of Independence, the Greek Army of Asia Minor had started their approach march in July 1921 and was advancing upon Anatolia. The Turks were preparing for the last life-and-death struggle east of Sakarya River. The Soviet Union was supposed to send artillery ammunition to Türkiye within the framework of the agreement signed on March 16, 1921. However, the Bolsheviks were afraid of the response from Western states and delayed this aid; moreover, they sent 77 mm artillery projectiles instead of the 75 mm Türkiye had requested. There was no time to correct the mistake because the Greek Army had approached the Sakarya positions.

There was great anxiety in the Ministry of National Defense, and solutions were sought to solve this wicked problem. Minister of National Defense, Refet Pasha held a meeting with personnel regarding the problem. Colonel Asım Bey, General Manager of War Manufactory, had brought the engineer Veli Bey to the meeting as chief of the armory department of the repair shop. After Refet Pasha explained the projectile problem, the young engineer Veli Bey offered to lathe the 77 mm projectiles directly without emptying their cartridges. Due to the danger of lathing full-round projectiles, no one had ever done it. Despite all the warnings, Veli Bey first built a small hand lathe, then took a projectile, successfully lathed the full-round projectile in a safe place, and was able to reduce its diameter by 2 mm. Next, the number of lathes was multiplied, and in a short time, enough ammunition had been provided to meet the needs of the battle.

In this example from the National Struggle, pragmatism resurfaced in the war as a limitation problem of philosophy. However, success in the National Struggle resulted from the will of a leader who, compared to his predecessors, acted with a clear purpose, wisely determined concrete objectives, and an unwavering belief that these could be achieved with science and technology.[12]

A problem similar to the one the Soviet Union had caused in the National Struggle was recently experienced with Israel, this time regarding uncrewed aerial systems (UASs). Since 2004, the PKK/KCK terrorist organization's actions had escalated again, and the need for UASs emerged. However, Türkiye did not receive the first delivery based on the agreement signed with Israel in 2001 until 2005. Moreover, the UAS supply was limited to reconnaissance UASs. Shooting at detected targets was a painstaking bureaucratic procedure. A-UASs, on the other hand, were never brought up. For this reason, although examples of successful operations had occurred, sufficient effectiveness could not be obtained from the UASs. Thereupon, Türkiye started to produce its own UAS technology and soon had one of the world's leading UAS fleets. Thus, if Türkiye is known as a drone power in international circles today, technology restrictions have had an undeniable role in this development.

The development of defense technologies is also seen to be closely dependent on field experience and feedback. In fact, the main strength of the discourse on domestic and national technology stems in practice from minimizing suspicion about mutual information sharing between the producer and the user. This has been experienced often in the field. Today, defense industry companies such as Aselsan, Havelsan, and Roketsan (Roket Sanayii ve Ticaret A.Ş.) are actively involved in areas of military operations. These companies have engineers in the field both to check the work of their subcontractors and to receive feedback from military troops (i.e., end-users) on its effectiveness. The troops can quickly maintain and repair the weapons and equipment they use, and the feedback regarding these processes can be sent to company headquarters rapidly through the field engineers. On the other hand, companies are also able to deliver new weapons and equipment they produce directly to the field, carrying out tests under real conditions. Figure 1.1 illustrates this symbiotic relationship.

In the symbiotic defense industry production mechanism, the producer and user work for a common purpose and are in direct contact with each other concerning the conflict; the instant feedback is taken seriously in the workshop and laboratory. This success of Türkiye's can be attributed to the pragmatic nature of Turkish strategic culture because purposefulness and utilitarianism are both based on concrete experiences and at the core of both the philosophy of technology as well as Turk's pragmatic strategic culture. For this reason, only decision-makers who are able to perceive earlier, understand the situation faster, make decisions quicker, and act sooner than the enemy can be successful in the complex and fuzzy operational environment of hybrid warfare. Such an ability requires well-educated pragmatic commanders as well as effective use of technology, and more so now than ever before.

The contemporary paradoxical situation between getting involved in a conflict and moving away from the tensions it creates is nothing new for decision-makers. What is new is that the paradox facing decision-makers is

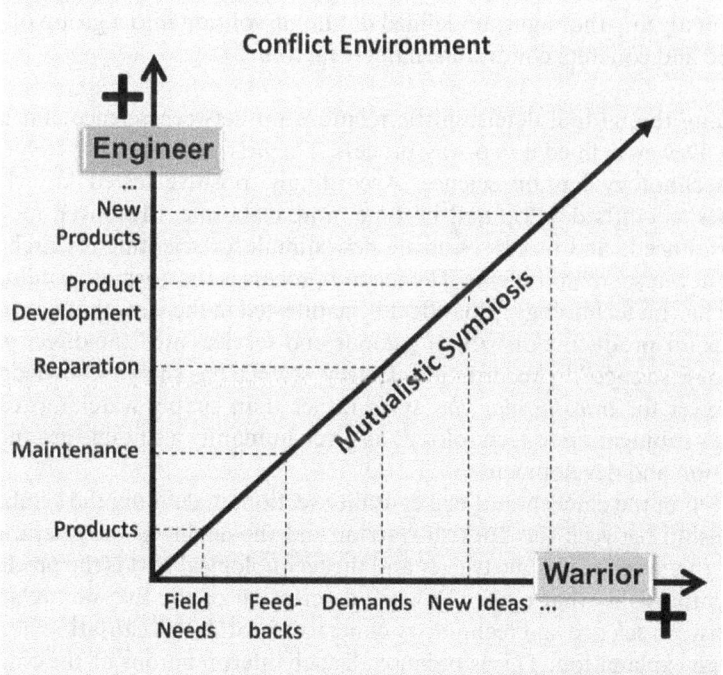

Figure 1.1 Türkiye's Symbiotic Defense Industry Production Mechanism.
Sources: *Created by the author.*

now rapidly deepening. That is why technology has been used to facilitate decision-making in the first quarter of the 21st century. The new multi-dimensional technologies constitute an essential part of increasingly robotic decision-making processes. Türkiye has had the chance to experience this path during operations since around 2012. In this context, new defense technologies have also become useful in Türkiye's counterterrorism operations in Iraq and Syria.

Fine-Tuning the Effect of Technological Determinism

The previous sections have explained the chicken-or-egg causation through the concept of symbiosis, but how can the symbiotic relationship between engineer and warrior affect institutional structure? The deterministic logic of Dewey's pragmatism works to explain this causality. According to Dewey, determinism is not a bad thing when used solely in the following sense:

> Much of the opposition to determinism is due, I believe, to the fact that the determinist either is understood to, or actually does, carry over into his use of the term "cause" this sense of efficient agency instead of using it in its sole justifiable scientific meaning of the analysis of a vague and unrelated fact into definite and cohering conditions. For my own part, I wish by "causation" to mean nothing more nor less than the possibility of analyzing the vague, undefined datum of volition into a group of specific and concrete conditions, namely factors.[13]

Regarding the mutual deterministic relationship between science and technology, Dewey defined a two-way process. The first process is the triggering effect technology has on science. Accordingly, positive tested knowledge increases as crafts develop and become more elaborate. Moreover, modern industrial needs and processes create new stimuli for scientific research and inspire new research problems. The second process is the normative influence science has on technology. This effect is manifested in the view that new technologies for producing/distributing goods and services are "the direct product of new science".[14] According to Dewey, science has shown knowledge to be a power for *transforming the world* rather than simply watching reality. Science's subjugation to technology has given humanity a suitable ground for innovation and development.

As seen in the chicken-and-egg causality section, in defining the symbiotic relationship between the Turkish warrior and the engineer, the operational part of this symbiosis is the trigger and the technological part is the producer. This symbiosis is what makes Dewey's description of the two-way relationship between science and technology come to life. Still, this causation needs a thorough explanation. This is because all such interpretations of the concept of technological determinism emphasize the importance of technology in social change for one reason or another, but how and why these interpretations differ is what makes technology so effective.

Habermas also sensed a problem here. According to him, industrial society has developed an overconfidence in the norms of efficiency, reason, and productivity in guiding this enterprise.[15] By adopting these as guidelines for making decisions about technology to the exclusion of other ethical criteria, society could produce a self-correcting process able to operate independently of the broader political and ethical contexts. This subsystem within society, composed of technologists who monitor the rationalization of life through the creation of technology, becomes autonomous once it frees itself from the ethical normative judgments of the public, or once society adopts its own standards of judgment.

Habermas's contribution to the definition of technological determinism is his focus on the norms of practice. Technology can be considered autonomous and decisive when the norms that drive its progress are removed from political and ethical discourse and replaced with value-based debates over efficiency or productivity goals, methods, alternatives, means, and ends. According to Jacques Ellul, technique is not just technology, it is the domination of social, political, and economic life together with accepted logic and efficiency goals.[16] As agents of value-laden norms and judgments, efficiency and technique lead to a technological society. A particular social practice and set of dominant norms produce this result. Lewis Mumford answered Habermas's question by calling attention to the dangers of *megatechnics*, which Marcuse describes as the one-dimensional man.

Cohen offers a different view of what constitutes technological determinism, which Bimber called the *logical sequence account*. Cohen's technological determinism views "machines and allied subhuman powers" as somewhat independent "agencies of history".[17] The argument here is that technology itself exerts a causal effect on social practice. Unlike Habermas' norm-based account, this account means that certain technological processes, regardless of their popularity or previous social practice, require organizational forms or political resource commitments once started. For example, tank production will require a suitable turret factory, a rubber and iron-steel factory that can meet the needs of the tank, and an electro-optical systems factory. According to this logically explained sequence, cultural and social change derive from technology. Technological developments take place according to a naturally given logic that is neither culturally nor socially determined, and these developments necessitate social adaptation and change.

The logical sequence account differs from Habermas' views on the production of a technological society. According to the norm-based account, what people begin to think about and desire is precisely what produces the technological society. However, according to the logical sequence account, self-propelled artillery comes after launched artillery, not by chance but because it is the next stage following a single great path of progress in the technical conquest of nature. As a result, the determinism of the logical sequence account is independent of culture and practice.[18] Determinism is the same everywhere. For example, the social implications of a technology like the musket in China are basically the same as in the Ottoman Empire, and the history of the musket must be critically tied to the same technical achievements in both cultures.

The unintended consequences account, sometimes referred to as technological determinism, is a third approach to explaining technology and social change. This account focuses on the undesirable effects of technological developments. The explanation derives from observations of the uncertainty and uncontrollability of the consequences of actions. For example, some politicians and generals in the USA believed that the use of nuclear weapons in 1945 would make the Japanese surrender and that the death and material and non-material destruction caused by the war would come to an end. However, the destruction caused by the nuclear bombs in Hiroshima and Nagasaki exceeded expectations. This dramatic, unintended consequence of the new technology came about unchecked. However, the unintended consequences account is problematic in terms of technological determinism. Winner pointed out that such accounts really amount to indeterminism as opposed to determinism.[19] Therefore, limiting this account to the disclosure of legal effects would be more appropriate.

In addition to Dewey's two-process (trigger and normative) deterministic pragmatic approach, the first two deterministic calculations have a share in defining Türkiye's national technology adventure. Essentially, Dewey's triggering process can be compared to Cohen's logical sequence account and the normative process to Habermas's norm-based account. Take UAS technology for example. Because limited UAS support from the USA had not reduced the number of casualties caused by terrorist attacks, a demand occurred in Turkish society for a UAS technology free from foreign dependency. This demand manifested itself as a Habermasian practice norm. Afterward, national UASs began being produced. This production, time independent of social demand, soon gave rise to equivalent, side, and supportive technologies just as in international technological trends; in other words, it developed a logical index.

As defense industry products find customers in the markets, economically these products will enable the development of subsidiary and auxiliary industries, which will in turn lead to the emergence of rival enterprises; the increased competition will then increase product quality and development. This economic impact is easily foreseeable. Yet the impact from technological development on the structure of the Turkish Armed Forces (TAF), which is the biggest customer of the defense industry, requires a more in-depth analysis. The effects of this process are observable. After experience gained in conflict areas, the technology demands of the TAF then mobilize many branches of the defense industry.

The rapid development of technology without waiting for doctrines to form on this subject, being able to produce technology at an appropriate amount, nurturing the production of technology based on field experiences, and the defense industry not being dependent on a uniform technology show the compatibility the defense industry has with the pragmatic strategic culture. Development is seen to not just result from demand in the field. The creative ideas of engineers and project entrepreneurs are also used once they are encountered in the field. This, in turn, develops new demands, new sectors, and sub-industries. As such, what is shown in Figure 1.1 can be expected to turn into a more parabolic view, as seen in Figure 1.2.

The new defense industry products have begun to change the way fighting occurs. The shaping of the battlefield has developed even more than before. Engagement rules are much more dynamic and are updated on a case-by-case basis. The size of the air threat has increased along with opportunities for air operations. Thanks to satellite technologies, communication methods are changing rapidly. The way to achieve maneuvering superiority is also changing. Moreover, with developing UAS capabilities, a new method known as uncrewed maneuvers has been added alongside manned maneuvers. Uncrewed maneuvers may be able to completely replace manned maneuvers in time. This gives soldiers the ability to track enemies without being in the same place, and to act with great precision. However, the psycho-social impacts of this on these soldiers are difficult to observe in the short term. On the other hand, this situation also changes commanders' decision-making processes and methods. Classical military decision-making processes are being replaced by ones that enable faster, more flexible decisions, and the

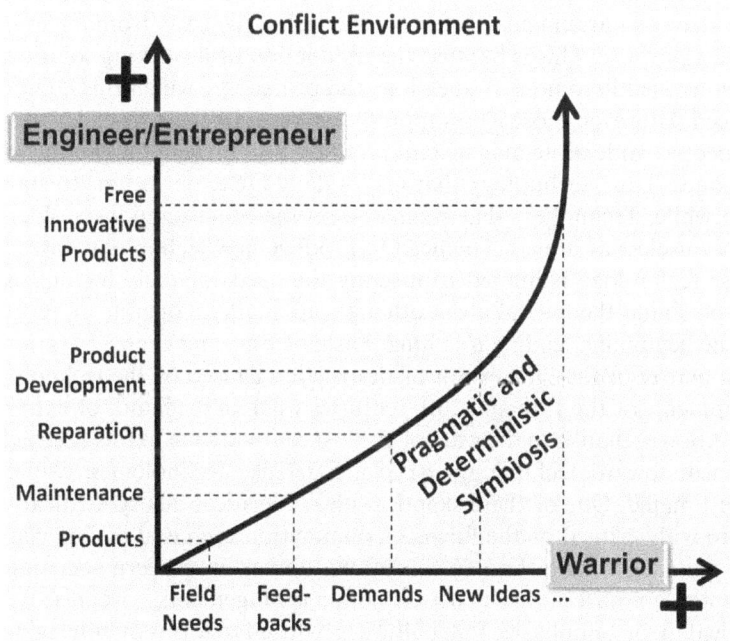

Figure 1.2 Türkiye's Pragmatic and Deterministic Defense Industry Production Mechanism.

Sources: *Created by the author.*

importance of operational design is increasing to avoid deficits in planning in a complex environment. The pragmatic fine-tuning nature of Turkish strategic culture may offer a new explanation in this regard. Fine-tuning behaviors that are able to yield valuable results may become the determinant finding of the TAF and produce national concepts and doctrines.

The Synergetic Mind of the Architect-Engineer is Superior to that of the Engineer

Meanwhile, technology's ability to produce solutions may create a conceptual trap. When technology is reduced to meeting field needs or limited to the relationship between the practitioner and the producer, the information produced with technology, that essentially enables the production of technology, can move away from *theoria* and become a process where daily needs are met. Isn't this what happened thousands of years ago with the Hittites and ancient Egypt? Then what is the problem? If the need has been identified or the purpose achieved, what is the purpose of the theory? A technique that is separated from science and reduced to meeting only daily needs loses its intellectual continuity and becomes limited to the time in which it is produced, or the object for which the need was met. For this reason, the technique in ancient Egypt was unable to exceed meeting the pharaohs' needs and solving some periodic security problems.

The strongest argument for moving the technique away from *theoria* is to achieve a purpose. This is because the production of a working solution will be seen as meddlesome if it questions the process by which this solution is produced. However, those who suppress the voices of these so-called meddlesome people while protecting instant products also prevent the production of better ones. Thus, technology that makes daily life easier begins to stagnate after a while. Technology that stagnates, on the other hand, gets older and becomes useless as it ages. The need to produce new technology arises. But, because technology is limited to meeting the needs of daily life and is not based on sound theoretical ideas, attempts are made to operate all processes from the beginning, leading to a huge waste of time and energy.

As a matter of fact, the extent of destruction caused by the technological developments of the 19th and 20th centuries when in the hands of extremists is now known, thanks to the wars of the last century. Due to this destruction, skepticism toward technology, especially defense technologies, cannot be underestimated. One of these skeptical ideas is critical theory. Critical theorists are truly critical of the forms of domination exercised by any class, or impersonal scientific-technological rational forces, in modern societies that have both capitalist and socialist economic arrangements.[20] Representatives of critical theory emphasize that culture is reduced to forms of entertainment that can be consumed massively by everyone through the *culture industry*, especially under capitalism. In other words, culture is commodified and marketed to the masses for profit. Critical theory reveals how individuals' consciousness is manipulated through mass culture, and how they are transformed into beings less inclined toward thinking and research by being seduced by consumption, especially in a modern society that has been reduced to a consumer society.[21]

For Heidegger, humanism grounded the essence of a human in the mind[22] and thus paved the way for humans to see those other than themselves as under their command. Heidegger criticized humanism, which he claimed has existed since the beginning of metaphysical thought and started with the discourse of *zoon echon logon*. Therefore, this definition of being human is what initiates technological thinking. The definition of *zoon echon logon*, which distinguishes people from other animals, also shows people superior to them. According to Heidegger, to think of a human in terms of animality, although not entirely wrong, is derived from metaphysical thinking.[23]

In this context, the criticisms brought by Heidegger's definitions of loss of existence, Marcuse's one-dimensionality, and Horkheimer's destruction of personal history have undeniably raised awareness among contemporary people. Martial technology, which can be called defense technology today, is what creates this doubt about technology. In fact, isn't the conceptual transformation of the war industry into a "defense industry" since the end of the 20th century also due to destructive and nefarious technology? Therefore, the risk of *dehumanization* needs to be added to the risk of *stagnation* emphasized in the previous paragraph. Otherwise, the use of defense industry products may be significantly restricted. So, instead of only producing a practical solution, there is greater value in determining the rules on how to use this practical solution, to make it permanent by theorizing, and thus inspire new practical solutions.

As is known, the economic situation in which a state that achieves a certain income level because of the advantages provided to it stagnates at this income level and is unable to get richer is called the middle-income trap. Inspired by the middle-income trap, the concept of *middle-technology trap* has been developed in recent years for technologies that falter after making certain progress.[24]

Although the symbiotic relationship between defense technologies and operational needs has worked for Türkiye, some risks are to be found in terms of the sustainability of this situation. Turkish defense industry products are updated versions of weapons and equipment previously used in Western countries in accordance with warriors' demands to benefit from domestic technological capacity. In other words, Türkiye produces current domestic versions of existing infantry rifles, self-moving arms, low altitude air defense weapons, tactical wheeled vehicles, thermal cameras, and UASs. Existing domestic systems that have been successfully tested at the operational level have medium value. While achievements at this level are encouraging, it is still too early to be complacent. There are very strong competitors, each of whom is a technology giant compared to Türkiye, and they should not be underestimated.

In order to turn the development of the Turkish defense industry into a true breakthrough, companies must be able to produce strategic, high-value, high-technology (high-tech) outputs such as uniquely designed engines, uniquely designed tanks, naval platforms, unique air/space vehicles, and innovative weapon systems that are unavailable in other states. Although some projects are encouraging in this regard, there is still a long way to go.

Nuclear technology takes the lead among strategic technologies today because it provides the most efficient energy and can produce the deadliest weapons. However, Türkiye is not expected to become a nuclear power. It has avoided nuclear weapons because of their inhumane destructive effects, pledging to be a conventional power from the very beginning.[25] For this reason, even more importance attaches to Türkiye's creation of non-nuclear high-tech products. Türkiye's ability to transfer the momentum it has gained in tactical- and operational-level defense technologies to the strategic level, without delay, will be what determines whether it can avoid the risk of stagnation.

Conclusion

The history of humanity can at the same time be called the history of technology. Technology is the most important and perhaps sole part of humanity's struggle to exist in nature. When mentioning technology, the first thing that comes to mind are the most advanced tools and equipment that facilitate human life at that time. However, no one questions the intellectual reasoning behind technology. Essentially, technology is not only a quality that separates humans from nature, but it is also the way in which humans interact with nature.

Technology without a clear purpose cannot be expected to have a beneficial effect. Therefore, the purposiveness of technology leads us to the operational utility argument. The concept of operational utility is the contribution pragmatists make to this field. Dewey focused on the concept of instrumentalism. According to Dewey's instrumentalist epistemology, humans interact with nature, and problem-solving is essential. The different ideas and methods used to solve problems are just the tools.

Dewey's views are valuable as they underpin the pragmatic nature of Türkiye's defense technologies, as Dewey is known to have come and stayed in Türkiye for about two months, to have met with Atatürk, and to have contributed significantly to the shaping of the Turkish education system. Therefore, the claim here is that pragmatism has the capacity to explain the causality effect of Turkish strategic culture on defense technologies.

On the other hand, pragmatic thinking needs to be accepted as having gained popularity in today's fuzzy strategic environment, one where crises have become routine, because in times of crisis, pragmatic solutions are needed even more. War, a limitation problem in political philosophy, appears plausibly to be closely related to pragmatism, for war by its very nature is a crisis.

Technological determinism is considered to be the most appropriate concept for explaining the development of the Turkish defense industry, an industry that inspires other branches of industry today. However, in any case, demand and feedback needs to be taken from the field and the uncontrolled effects need to be considered. Meanwhile, the deterministic course of technology has led to changes in the Turkish strategic culture and the way war occurs in this context. Success is being able to manage this change and produce information in a way that is useful. The most concrete manifestation of this fine-tuning behavior may be the production of forward thinking and original concepts and doctrines.

Turks owe the practical and useful solutions they produce in times of crisis to their pragmatic cultural qualities. This is also because the geographical environment they live in is in a state of constant crisis. For this reason, the efforts they make while overcoming the crises they face also benefit technological development. In recent years, the production-demand relationship between defense industry companies and military forces has gained a symbiotic nature. Today, the Turkish defense industry has become competitive in many fields in world markets. However, turning the development into a breakthrough requires original designs that are able to affect the strategic level. Therefore, producing high-tech conventional strategic systems should be essential, and a priority for Türkiye, as it has made the conscious choice to abandon nuclear weapons technology. Although positive developments exist in this regard, they should overcome the obstacles, travel the long path in front of them, and achieve a level of contemporary civilization with perseverance, and without lethargy.

Notes

1 Frank G. Hoffman, *Conflict in the 21st Century: The Rise of Hybrid Wars*, Potomac Institute for Policy Studies, Virginia, 2007, p. 52.
2 John Dewey, *Reconstruction in Philosophy*, Henry Holt & Company, New York, 1920, p. 75, 178.
3 Dewey, *The Quest for Certainty: A Study of the Relation of Knowledge and Action*, Balch & Company, New York, 1929, pp. 103–106.
4 Hikmet Kırık and Oya Morva, *Cumhuriyet ve Pragmatizm: Yazılmamış Kuram*, Doruk Yayınları, İstanbul, 2016, p. 29.
5 While in Istanbul, Dewey presented his first report on education, which included resources that urgently needed to be allocated from the budget. He sent his main report after he had returned to the USA.
6 Bahri Ata, "1924 Türk Basını Işığında Amerikalı Eğitimci John Dewey'nin Türkiye Seyahati", *G. Ü. Gazi Eğitim Fakültesi Dergisi*, 21:3, 2001, pp. 193–207.
7 Dewey, *Türkiye Maarifi Hakkında Rapor*, Devlet Matbaası, İstanbul, 1939. TBMM Kütüphanesi Koleksiyonu, https://acikerisim.tbmm.gov.tr/xmlui/bitstream/handle/11543/928/197000571.pdf?sequence=1&isAllowed=y, [accessed on 22 August, 2021].
8 Kırık and Morva, *Cumhuriyet ve Pragmatizm*, pp. 130–143.
9 Sun Tzu, *The Art of War*, Trans. Lionel Giles, Aziloth Books, Durham, 2010, p. 6 (III/2).
10 It is expressed as "troop-made" in the common jargon of the Turkish Army.
11 Ersoy Zengin, "Millî Mücadele Yıllarında İmalat-ı Harbiye Fabrikaları", *Mavi Atlas*, 5:1, 2017, pp. 201–223; Turgut Özakman, *Şu Çılgın Türkler*, Bilgi Yayınları, Ankara, 2005.

12 Kırık and Morva, *Cumhuriyet ve Pragmatizm*, p. 93.

13 Dewey, "The Ego as Cause," *Philosophical Review*, 3, 1894, pp. 337–341, p. 337.

14 Dewey, *Logic, the Theory of Inquiry*, Henry Holt & Company, New York, 1939, pp. 74–75.

15 Dewey, *Logic, the Theory of Inquiry*, Henry Holt & Company, New York, 1939, pp. 93.

16 Jacques Ellul, *The Technological Society*, Vintage, New York, 1964, pp. xxvi, 137, 153.

17 Gerald A. Cohen, *Karl Marx's Theory of History: A Defence*, Princeton University Press, 1978, p. 146.

18 Robert Heilbroner, "Do Machines Make History?", *Technology and Culture*, 2, Winter 1961, p. 337.

19 Langdon Winner, *Autonomous Technology*, The MIT Press, Cambridge, 1977, p. 42.

20 Martin Slattery, *Key Ideas in Sociology*, Nelson Thornes, Cheltenham, 1991, p. 106.

21 Slattery, *Key Ideas in Sociology*, p. 107.

22 Martin Heidegger, "Letter on Humanism", in (ed.) William McNeil, *Pathmarks*, trans. Frank A. Capuzzi, Cambridge University Press, Cambridge, 1998, p. 246.

23 Heidegger, *Parmenides*, Trans. Andre Schuwer & Richard Rojcewicz, Indiana University Press, 1998, p. 52.

24 Antonio Andreoni, Fiona Tregenna, "Escaping the middle-income technology trap: A comparative analysis of industrial policies in China, Brazil and South Africa," *Structural Change and Economic Dynamics*, 54, 2020, pp. 324–340; İbrahim S. Akçomak and Serkan Bürken, "Middle-Technology Trap: The Case of Automotive Industry in Turkey," in (eds.) Ferreira, J.J.M., Teixeira, S.J., Rammal, H.G., *Technological Innovation and International Competitiveness for Business Growth*, Palgrave Macmillan, Cham, 2021.

25 Türkiye is party to the main international disarmament and non-proliferation treaties and regimes. These are: Treaty on Non-Proliferation of Nuclear Weapons (NPT); Review Conferences (RevCon); Comprehensive Test Ban Treaty (CTBT); Chemical Weapons Convention (CWC); Biological and Toxin Weapons Convention (BWC); The Convention on the Prohibition of the Use, Stockpiling, Production and Transfer of Anti-Personnel Mines and on their Destruction (Ottawa Convention); Convention on Prohibitions or Restrictions on the Use of Certain Conventional Weapons (CCW) (Protocol I, Amended Protocol II and Protocol IV); Hague Code of Conduct against Ballistic Missile Proliferation (HCOC); Ministry of Foreign Affairs Site, "Arms Control and Disarmament," https://www.mfa.gov.tr/arms-control-and-disarmament.en.mfa, accessed on 28.08.2021.

Bibliography

Akçomak, İbrahim S., and Serkan Bürken, "Middle – Technology Trap: The Case of Automotive Industry in Turkey", in (eds.) Ferreira, J.J.M., Teixeira, S.J., and Rammal, H.G., *Technological Innovation and International Competitiveness for Business Growth*, Palgrave Macmillan, Cham, 2021, pp. 263–306.

Andreoni, Antonio, and Fiona Tregenna, "Escaping the middle – income technology trap: A comparative analysis of industrial policies in China, Brazil and South Africa", *Structural Change and Economic Dynamics*, 54, 2020, pp. 324–340.

Ata, Bahri, "1924 Türk Basını Işığında Amerikalı Eğitimci John Dewey'nin Türkiye Seyahati", *G. Ü. Gazi Eğitim Fakültesi Dergisi*, 21:3, 2001, pp. 193–207.

Cohen, Gerald A., *Karl Marx's Theory of History: A Defence*, Princeton University Press, Princeton, 1978.

Dewey, John, "The Ego as Cause", *Philosophical Review*, 3, 1894, pp. 337–341.

Dewey, John, *Reconstruction in Philosophy*, Henry Holt & Company, New York, 1920.

Dewey, John, *The Quest for Certainty: A Study of the Relation of Knowledge and Action*, Minton, Balch & Company, New York, 1929.

Dewey, John, *Logic, the Theory of Inquiry*, Henry Holt & Company, New York, 1939a.

Dewey, John, *Türkiye Maarifi Hakkında Rapor*, Devlet Matbaası, İstanbul, 1939b. TBMM Kütüphanesi Açık Erişim Koleksiyonu, https://acikerisim.tbmm.gov.tr/xmlui/bitstream/handle/11543/928/197000571.pdf?sequence=1&isAllowed=y.

Ellul, Jacques, *The Technological Society*, Vintage, New York, 1964.

Heidegger, Martin, "Letter on Humanism," in (ed.) William McNeil, *Pathmarks*, trans. Frank A. Capuzzi, Cambridge University Press, Cambridge, 1998a.

Heidegger, Martin, *Parmenides*, Trans. Andre Schuwer and Richard Rojcewicz, Indiana University Press, Indianapolis, 1998b.

Heilbroner, Robert, "Do Machines Make History?", *Technology and Culture*, 2, Winter 1961, pp. 335–345.

Hoffman, Frank G., *Conflict in the 21st Century: The Rise of Hybrid Wars*, Potomac Institute for Policy Studies, Virginia, 2007.

Kırık, Hikmet, and Oya Morva, *Cumhuriyet ve Pragmatizm: Yazılmamış Kuram*, Doruk Yayınları, İstanbul, 2016.

Ministry of Foreign Affairs Site, (n.d.) "Arms Control and Disarmament", https://www.mfa.gov.tr/arms-control-and-disarmament.en.mfa.

Özakman, Turgut, *Şu Çılgın Türkler*, Bilgi Yayınları, Ankara, 2005.

Slattery, Martin, *Key Ideas in Sociology*, Nelson Thornes, Cheltenham, 1991.

Tzu, Sun, *The Art of War*, Trans. Lionel Giles, Aziloth Books, Durham, 2010.

Winner, Langdon, *Autonomous Technology*, The MIT Press, Cambridge, 1977.

Zengin, Ersoy, "Millî Mücadele Yıllarında İmalat-ı Harbiye," *Mavi Atlas*, 5:1, 2017, pp. 201–223.

2 The Turkish Defense Industry in the Post-Cold War Era

Bound to Emerge

Tolga Öz

Introduction

The greatest reflections to land on the pages of recorded human history have been about the wars that opened and closed eras and removed many great states and empires from the stage of history. Humanity has been able to live in peace for only 268 of the last 3,400 years.[1] This situation has revealed that war may occur at any time, and even if it does not, countries should always be prepared for its possibility. States have raised armies to protect their survival, provide deterrence against their current or potential opponents, resist attacks, and achieve national interests. Providing defense services has necessitated making defense expenditures by allocating a budget from the nation's income at the expense of its welfare. Several critical turning points have impacted the portion allocated to defense matters, a recent turning point being the end of the Cold War.

The end of the Cold War created an optimistic attitude toward the possibility of reducing military spending in favor of other public expenses, particularly in Europe. With regard to how money is spent, a clear distinction has been made between NATO and the defunct Warsaw Pact countries, and the rest of the globe.[2] The main actors have reduced their armies in size and number, changing their equipment and force structures toward mobile, flexible, and modular forces. However, regional conflicts have begun to emerge in different parts of the world. As a result, world military expenditures have increased by 3.6% compared to 2018, reaching $1.917 trillion USD in 2019 as the highest jump of the last decade and a historical peak.[3] Moreover, the 2.6% increase in world military spending came in a year when the global gross domestic product (GDP) had shrunk by 4.4%, primarily due to the economic impacts of the COVID-19 pandemic. As a result, military spending as a share of GDP (i.e., the military burden) reached a global average of 2.4% in 2020, up from 2.2% in 2019. This was the most significant year-to-year rise in the military burden since the global financial and economic crisis of 2009.[4] When evaluated on a regional basis, one would not be wrong to say that the continued global arms race is one of the main driving forces of the increase in armed conflicts, and the volatile nature of defense and military expenditures in all regions of the world, including sub-Saharan Africa. Additionally,

DOI: 10.4324/9781003327127-3

technological developments have facilitated countries' ability to notice and compare the weaknesses and strengths of their defense industries with other countries. This increased awareness also generally triggers growth in defense expenditures.

Throughout the 1990s, Türkiye faced various significant challenges to its national interests, territorial integrity, and economic well-being. In contrast, other NATO nations that had been liberated from the Soviet menace did not have such terrorism problems. Separatist terrorism has been the main difference between Türkiye and other NATO countries. These threats require the sustainment of a broad set of military capabilities to deter and, if necessary, fight against various opponents. Accordingly, maintaining this military power and using it as needed correlates to a robust defense industry. The divergence of Türkiye's threat perception from that of other NATO countries is one of the factors that has driven Türkiye to gain an indigenous and self-sustainable defense industry. The lessons learned from the past embargoes and sanctions of allied countries have also helped Turks invest in defense technologies and related sectors as well as in capable, well-educated experts.

This research investigates Türkiye's efforts to establish an indigenous, self-sufficient defense industry that is able to fulfill the needs of the Turkish Armed Forces (TAF) while remaining competitive in a rapidly changing marketplace. It aims to uncover the evolution of the transformation process of the Turkish defense industry along with the major strategies that have been adopted through the critical achievements in developing a competitive defense industry based on Türkiye's ever-changing security policy and risk perception.

Defense expenditure is classified as the part of the consolidated budget reserved for defense purposes to ensure a country's survival at home and abroad.[5] In more detail, defense expenditure refers to expenditures on personnel, the purchase of goods and services (e.g., weapons, vehicles, equipment, maintenance, repair, construction, research and design or R&D) acquired for use in a country's field of defense, and expenditures on defense aid and deployment.[6] In other words, defense expenditure guarantees nations' survivability and sovereignty.[7]

Research on defense expenditure has suggested that no global consensus exists regarding how to classify defense expenditures. Yet defense expenditures are one of the main components of public spending. The defense burden of countries varies depending on the economic, social, and political dimensions of both the local and international environments. As such, international institutions have shown different classifications for defense expenditures.[8] In other words, states have different perceptions and assessments regarding how defense expenditures are classified because of the differences in their social, demographic, and economic structures as well as threat assessments. For this reason, no specific evaluation can be made when comparing states' defense expenditures.[9] For example, the budget of an expenditure item within one country's ministry of defense may be included in another country's budget for their ministry of health. Various data sets have reflected this difference. Consequently, the Stockholm International Peace Research Institute (SIPRI), the International Institute for Strategic Studies (IISS), Military Balance, and lastly NATO all have different defense expenditure definitions reflected in their statistics. These variants may cause contrasting analyses; however, long-term perspectives might be helpful for overcoming this difficulty. Therefore, this study focuses on the last 30 years of Turkish defense expenditures by providing a general outline of its evolution. Of course, the critical turning points of the past will additionally be examined in order to understand the emphasis on having domestic and nationally-produced weapon systems and capabilities.

Factors Affecting Türkiye's Defense Expenditures

The factors affecting defense expenditures vary according to different countries' threat perceptions. As for Türkiye, its geographical location and deterrence factors come first in determining the defense expenditure budget. Other factors are participation in international political and military organizations, the defense expenditures of neighboring countries, domestic politics, and economic and technological developments.[10]

Türkiye is located in a critically important geographical region where wars have arisen for thousands of years for a variety of reasons, with armed conflicts continuing to occur even today. The fact that Türkiye is surrounded by seas on three sides and has complete control of passage to and from the Black Sea by means of the Turkish Straits increases Türkiye's geopolitical importance through significance of these topographical features in world trade. The evident fact that the military balance in the Black Sea has shifted rather significantly in Russia's favor since the seizure of Crimea and the further militarization of the peninsula should additionally be taken into consideration. Alongside this, the recently discovered rich hydrocarbon reserves both in the eastern Mediterranean and the western Black Sea's exclusive economic zones have become the focus of new emerging economic energy opportunities attracting the attention of the European Union (EU) and the rest of the world. This is because the EU currently has the second largest economy in the world in terms of purchasing power parity after the USA, and imports 90% of its oil and 66% of its natural gas. When considering that the EU's energy dependency and needs will increase in the future, the EU will turn to alternatives other than Russia, which already meets 30% of the Union's natural gas supply.

Türkiye's unique location, growing market economy, long collaboration with the West and NATO, and possession of the second largest armed forces make her a geostrategic player in world affairs. This geopolitically important area, the heart of Eurasia, has always been the center of world politics. Türkiye is centrally located between Europe and Asia, where most of the world's political and economic power is concentrated and has recently become a vital center of economic growth and increased political influence.[11]

Therefore, substantial developments in this geographical setting are a considerable matter that will directly affect the amount of defense expenditure in the context of Türkiye's foreign policy. Based on these facts and its geopolitical position, great importance is found in Türkiye having an effective, deterrable, respected, and operationally advanced armed forces with a cost-effective, source-based, sustainable defense industry.

States participate in multinational military organizations in order to partner with different countries. Each state gains the advantage of reducing security-related expenses by participating in international organizations. In other words, defending the unity becomes a priority for member countries who protect each other. This reduces defense expenditure costs and increases the benefit from the expenditure the other countries make. Türkiye follows NATO guidelines; however, terrorism threats, regional conflicts, and economic restraints have not always allowed it to reach the NATO average. The total defense expenditure of NATO member countries in 2018 was approximately $963.1 billion USD. In the same year, the total defense expenditure of other countries in the world was $819.2 billion USD.[12] The 2014 NATO Wales Summit reached the agreement that member countries should spend at least 2% of their GDP on defense and allocate at least 20% of this expenditure to R&D and system procurement. Member countries that had not yet reached these levels were expected to achieve their targets within ten years. With a 1.86% share, Türkiye is close to reaching this NATO goal. In addition, Figure 2.1. shows that Türkiye allocates 38.6% of this share to R&D, well above the set target.

Another factor that should be considered is neighboring countries' defense investments. The reactionary expenditures made as a result of a neighboring country's defense expenditure cause the relevant country to engage in an armament race.[14] An example of this was the USSR's arms race with the USA during the Cold War. According to SIPRI data, another example of this is China, Pakistan, and India as bordering countries realizing the highest defense expenditures in 2013.[15] In addition, the speed with which armaments have arisen from various problems between Türkiye and Greece can be given as yet another example. Table 2.1. provides the annual defense expenditures of Türkiye and its neighboring countries and the ratios to their respective GDPs.

Lastly, developing countries generally cannot meet their own defense needs compared to developed countries simply because they cannot establish a sufficient defense industry, causing them to allocate a higher share of their national income to defense expenditures than developed countries do.[16] In this context, arms export sales by both private and state-owned defense industry companies contribute significantly to the national income. In this regard, Turkish defense sector firms have started to establish a foothold in the global rankings, thus contributing to the Turkish economy. For example, Aselsan is one of the three Turkish companies on the SIPRI Top 100 list with $1.740 billion US in sales in 2018 and ranking 54th, while Turkish Aerospace

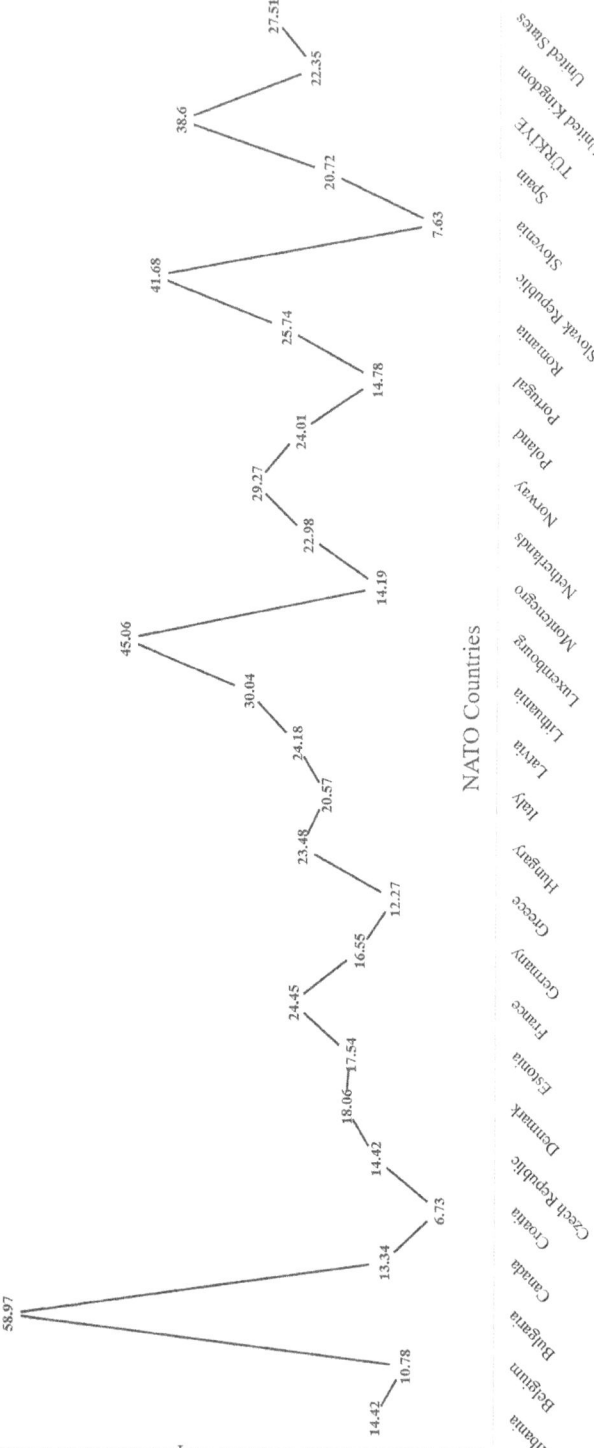

Figure 2.1 2019 Annual Share of NATO Countries' Equipment and R&D as a Percentage of Defense Expenditure.

Sources: *Created by the author. Source: Defence Expenditure of NATO Countries, 2019.*[13]

Table 2.1 Defense Expenditures of Türkiye and Neighboring Countries[1]

Year	Spending Ratio/ Amount	Türkiye	Greece	Iran	Armenia	Iraq	Georgia	Bulgaria
2020	Million US$	**11,038**	4,984	14,059	621	10,267	288	1,158
	% of GDP	**1.7**	2.56	2.3	4.85	5.76	1.89	1.7
1990	Million US$	**1,600**	2,209	3,810	652	4,799	311	1,793
	% of GDP	**2.6**	5.8	3.3	4.8	unavail.	1.7	3.5

Created by the author. Source: The IISS, The Military Balance 2021 and (1991–92) issues.

1 Data compiled from The International Institute for Strategic Studies (IISS), The Military Balance 2021, Routledge (2021) and (1991–92) issues.

Industries (TUSAŞ) came in 84th with $1.7 billion US in sales.[17] Roketsan A.Ş. supplies the NATO Support and Procurement Agency (NSPA), which provides NATO's integrated logistics and services[18], making $1.740 billion US in sales and ranking 89th on the 2019 SIPRI Top 100 list of world defense industry actors, rising up from 96th in 2018. In addition, BMC Automotive Industry and Trade Inc. was included in the 2019 Defense News list for the first time at 85th place with $1.74 billion US in sales.[19]

Evolution of the Turkish Defense Industry

Transforming the TAF into a modern army possessing advanced technological weapons and equipment has been one of the most important goals since the first years of the Republic. However, both Türkiye's geography and underdeveloped economy have prevented this goal from being fully achieved. Thus, despite being considered the second largest army of NATO in terms of size, TAF can safely be said to have actually been a low-tech army. One striking example regarding this subject is seen in the following speech, which Birand quoted from a US Congress session in 1982:

Almost all the basic weapons of the Turkish Army have become obsolete. These include not only tanks, ships, and aircraft but also communications and support units. Almost all of the 90 mm tank guns are ineffective, and since their engines operate on gasoline, they have limited range… Only 1% of anti-tank guns are modern and effective. 89% of short-range air defense weapons are from the 1940s, 93% of FM communication radios are unusable… All destroyers in the Navy are the American ships produced in World War II era. 75% of submarines are 35-year-old boats that have reached the end of their life… In the Air Force, 70% of fighters are from the 1960s, and they have none of the modern defense systems required to protect airports…[20]

The above example shows how TAF had significant problems obtaining new weaponry and equipment. Overcoming the problems caused by inadequate and outdated weapons and equipment is not easy, even in the fight against the internal security problems that reached their peak in the early 1990s. The First Gulf War in 1991 also increased pressure on the political and military leaders regarding TAF not having equipment or weapons systems capable of eliminating the threats emerging in the post-Cold War era.[21]

One of the most prominent issues in evaluating Türkiye's military is the size of its defense budget. Türkiye having a higher-than-needed defense budget has been widely debated. Indeed, at first glance the share allocated to the army shows that the necessary condition for creating a developed army has been met. According to World Bank data, the ratio of Türkiye's defense budget to GDP, which was 2.94% in 1988, reached 4.14% in 1996 then declined to 2.3% in 2012. According to another source, while the ratio of the defense budgets of NATO countries to GDP was 2.76% in 1990, this ratio was 2.6% in Türkiye. Even the European NATO average exceeded Türkiye's at 2.63% for the same year. By 2013, the NATO average had decreased to 1.38%, with Türkiye's defense budget ratio being 1.26% for the same period. Therefore, in parallel with other NATO countries, Türkiye had experienced a 50% decrease. In 2020, this ratio was 1.86% and ranked 12th within NATO. All these figures may convince analysts that the economic support required for TAF to become an effective and modern army has been sufficiently provided. However, evaluating the size of the defense budget in terms of GDP alone is not sufficient to understand the actual situation. Statistics such as *technological sophistication* and *defense expenditures per capita* should also be considered. For TAF, these criteria reveal a very different picture to that shown by the share allocated from the GDP.

The technological sophistication index figure is obtained by dividing the defense budget by the number of soldiers.[22] Table 2.2 shows the technological development figures of NATO members in the post-Cold War period. Accordingly, this figure for Türkiye is seen to have increased approximately eleven-fold in this period. Meanwhile, another important indicator is defense expenditure per capita. This figure rose from a deficient level of US $29 in 1990 to US $135 in 2020. However, a comparison within NATO demonstrates this improvement to be insufficient as Türkiye ranks among the bottom. This is mostly due to the presence of a conscripted army, which may shadow the real situation on the ground, in which only professional units operate and therefore have the biggest share. Additionally, one should also explore Türkiye's past to gain deeper insight into its current military capabilities and defense industry.

Table 2.2 A Comparison of the Defense Statistics for 1990 and 2020[1]

Country	# of Military Personnel in 1990 (x 1000)	# of Military Personnel in 2020 (x 1000)	1990 Defense Budget in Millions USD	2020 Defense Budget in Millions USD	1990 Defense Budget GDP %	2020 Defense Budget; GDP %	Technological Development 1990 (USD)	Technological Development 2020 (USD)	1990 Defense Spending per capita (USD)	2020 Defense Spending per capita (USD)
USA	2,029	1,388	249,149	738,000	5.4	3.55	122,758	531,700	1,001	2,219
UK	300	149	19,574	61,526	3.7	2.33	65,225	412,926	346	936
Norway	33	23	1,880	6,493	3.3	1.77	57,492	282,304	448	1,188
Netherlands	101	34	4,134	12,615	2.8	1.42	40,769	371,029	280	730
Denmark	29	15	1,253	4,909	2	1.45	42,619	327,266	246	836
Luxembourg	0.8	1.1	52	390	1.3	0.57	65,000	354,545	142	621
Canada	87	67	7,064	20,049	1.7	1.25	81,570	299,238	265	532
Germany	476	184	16,940	51,347	2.2	1.36	35,566	279,059	281	641
France	453	203	18,113	55,034	2.8	2.16	39,975	271,103	321	811
Belgium	85	26	1,558	5,453	1.6	1.08	18,232	209,730	158	465
Italy	361	166	9,320	29,344	1.8	1.59	25,789	176,771	163	470
Poland	305	114	2,226	12,875	1.2	1.22	7,298	112,938	58	336
Czech Republic	154	25	4,059	3,278	3.7	1.35	26,357	131,120	259	306
Spain	257	123	3,742	12,985	1.8	1.04	14,537	105,569	94	260
Portugal	62	27	753	2,857	2.6	1.29	12,184	105,814	72	277
Bulgaria	107	37	1,793	1,158	3.5	1.70	16,757	31,297	198	166
Hungary	86	23	1,149	2,041	2.1	1.36	13,283	72,892	109	209
Greece	159	143	2,209	4,984	5.8	2.56	13,937	34,853	218	470
Romania	200	69	1,287	5,207	1.85	2.09	6,409	75,463	55	244
Türkiye	**579**	**355**	**1,600**	**11,038**	**2.6**	**1.70**	**2,762**	**31,092**	**29**	**135**

(Continued)

Table 1.2 (Continued)

Country	# of Military Personnel in 1990 (x 1000)	# of Military Personnel in 2020 (x 1000)	1990 Defense Budget in Millions USD	2020 Defense Budget in Millions USD	1990 Defense Budget GDP%	2020 Defense Budget; GDP%	Technological Development 1990 (USD)	Technological Development 2020 (USD)	1990 Defense Spending per capita (USD)	2020 Defense Spending per capita (USD)
Albania	48	8	114	187	4.2	1.33	2,375	23,375	35	61
Macedonia, North		8		188		1,51		23,500		89
Russia	5,300	900	91,631	43,184	11.1	2.95	17,289	47,982	318	305

Created by the author. Source: The IISS, The Military Balance 2021 and (1991–92) issues.

1 Data compiled from The International Institute for Strategic Studies (IISS), The Military Balance 2021 and 1991–92 issues. An earlier version of this compilation can be found in B.Ateş, *Soğuk Savaş Sonrası Dönemde Askeri Değişim: NATO Orduları ve Türk Silahlı Kuvvetleri Üzerine Karşılaştırmalı Bir Analiz*, Unpublished doctoral dissertation, Gazi University, Ankara, 2014, p. 120.

The Initial Attempts to Build a National Defense Industry

After the long-lasting Independence War of the young Turkish Republic, almost no tangible heritage or infrastructure had been carried over from the Ottoman Empire on behalf of the industry, specifically the defense industry, apart from a few armor production plants. However, beginning with the early years of the Republic (1923–1950), the defense industry was identified and specified as a premier theme on the national priority list for the state's development of a total national industry. As a result, a state-supported defense industry strategy was first implemented at the Izmir Economic Congress in 1923, with Ataturk pinpointing the importance of the defense industry. Moreover, government encouragement and support for the establishment of new institutions and factories (e.g., some small arms and repair shops, and cartridge factories) in different industrial defense areas were also underlined in the same congress despite poor economic, technological, and infrastructural conditions, as well as inexperienced and inadequate human resources.

In the interwar period, the agenda aimed to implement joint state and private sector projects for developing the domestic defense industry. However, some of these failed due to licensing problems and bureaucratic obstacles, while others were unsuccessful due to free foreign military aid. For example, in 1926, Tayyare and Motor Türk A.Ş. (TOMTAŞ), an aircraft factory partnered with German firm Junkers, was established in Kayseri and achieved a 112-aircraft production capacity in 1939 in the aviation industry. However, due to a so-called license conflict with the Germans, this promising aircraft factory later had tragically to be shut down. Apart from these types of Turkish industry defense entrepreneurships for meeting the requirements of internal security under the guidance of the state, other critical international relations-based improvements can be addressed. One of these is the Lend-Lease Act, dating back to March 11, 1941. Under the scope of this act, the United States provided weapon transfers and ammunition support to various countries such as Britain, China, the Soviet Union, France, and Türkiye. As part of the program, USA supplied war equipment worth $95 million US to Türkiye between 1941 and 1944.[23]

The Truman Doctrine from July 12, 1947 was also a critical milestone in Türkiye's defense industry evolution. The Truman Doctrine was the primary provision for Türkiye until the 2000s, not only for support and aid packages but also for requesting and using credits from Western economic institutions, especially from the USA. Respectively, as a new member of the UN in 1945, the European Council in 1949, and NATO in 1952, as well as the Ankara Agreement in 1963 and its European Union membership application in 1987, Türkiye's puzzling situation increased after World War II not only through its military but also its economic and political dependency. Moreover, in the bipolar world of those days as a NATO member country, the Soviet threat was another reality that should be considered a catalyst for Türkiye receiving more military aid from the USA.[24] Consequently, the surplus of goods and equipment granted to Türkiye also prevented domestic production and sub-jected Türkiye to NATO markets.

Lessons Learned from the Embargoes

During the Cold War, Türkiye's defense industry strategy based on foreign aid and supply caused capacity and production losses in domestic military facto-ries. In the end, this strategy resulted in these factories transforming in 1950 into the basis for the General Directorate of the Machinery and Chemical Industry Corporation (MKEK) as a state-owned enterprise. However, this enterprise lacked the ability to support the Turkish military, as was proven during the Cyprus Peace Operation in 1974 when the US imposed an arms embargo on Türkiye. The US imposed embargoes by totally cutting off Truman Doctrine grants, as well as Foreign Military Financing (FMF) and Foreign Military Sales (FMS) credits, commercial credits, and training sup-port between 1975 and 1980. With the termination of embargoes at the begin-ning of the 1980s, Türkiye and the USA signed the Defense and Economic Cooperation Agreement, which reinitialized the FMF and FMS budgets until 1998.[25] This embargo was one of the strongest motivations for building a self-sufficient domestic defense industry. As a result, Turkish industrialization gained momentum beginning in the 1970s. This awakening process supported the establishment of diverse industries and even promoted a project for man-ufacturing a national military tank, despite lacking any infrastructure, know-how, R&D capacity, or financial resources. In fact, one of these organizations was Türkiye's premier tractor and diesel engine producer, TÜMOSAN (est. 1975), which today provides support for the Altay Tank Project.

After ending the official sanctions against the Cyprus Peace Operation in the 1980s, Türkiye suffered from implied and indirect embargoes regarding various incidents, especially during counterterrorism operations to neutralize PKK during the 1990s. For example, after Türkiye agreed to purchase the AH-1W Super Cobra Helicopter from the US, this FMF & FMS request endorsement was rejected by the US administration and US Congress due to certain lobbying activities and political issues. Correspondingly, US Congress again refused to grant the purchase of Reaper drones and Cobra helicopters during the 2000s. In 2015, US Congress again rejected Türkiye's request to purchase excess US Navy guided missile frigates, instead agreeing to export them to non-NATO member countries Mexico and Taiwan.[26] In addition to the US sanctions, arms sales, and the export of related equipment from coun-tries such as Germany, Sweden, and Canada were halted under the pretext of human rights. All these restrictions necessitated Türkiye establishing a domestic defense industry through its own means.

The Emergence of a National Defense Industry

During the 1980s, Türkiye embraced the autarkic and national weapon pro-duction strategy as an outcome of increased awareness concerning the importance of defense industry reforms and capabilities and began experimenting with structural reforms. In order to reduce foreign dependency in the defense industry, investments were restarted with the founding of the Turkish Aircraft Industries Corporation (TAI) under the Ministry of Industry and Technology in 1973, which was followed by the establishment of companies such as Aselsan (1975), Isbir Holding (1978), Aspilsan Inc. (1981), and Havelsan (1982) by the Land, Naval, and Air Force Support Foundations. Ultimately, all these establishments were united under the Turkish Armed Forces Foundation (TAFF) in 1987 as one of the current prominent actors of the defense industry.[27] Even TAI, one of TAFF's defense business enterprises, became the first Turkish subcontractor firm for the F-16 Fighting Falcon Fighter Aircraft project.

Another prominent actor of the same period, the Defense Industry Development and Support Administration Office (SaGeB), was founded in 1985 under the Ministry of National Defense per Law No. 3238. SaGeB's tasks were to set policies regarding establishing the infrastructure of the defense industry with the authority and responsibility to implement these policies. Subsequently, SaGeB would be restructured as the Undersecretariat for Defense Industries (SSM) in 1989. It became affiliated with the Presidency of the Turkish Republic in December 2017 and was renamed the Presidency of the Defense Industry (SSB) in July 2018 in order to develop a national modern defense industry for modernizing the Turkish Armed Forces.[28] Furthermore, in accordance with the aims and policies contained in Law No. 3238, the foreign procurement strategy between 1990 and 2000 changed from direct procurement to co-production.

Following these reforms and with the suggestions and support of the Ministry of National Defense, the Defense Industry Manufacturers Association (SASAD) was established in Ankara in 1990. In the beginning of 2012, civil aviation and space manufacturers also joined the association. As a result, the association was renamed the Defense and Aerospace Industry Manufacturers Association, with 122 affiliated members as of 2020.[29] Furthermore, the foundation of another premier actor in the Turkish defense industry, Defense Technology, Engineering, and Trade Inc. (STM) in 1991 was also a breakthrough, not only for the Turkish but also for the global defense industry, with its smart, innovative, and competitive technological solutions reaching 92nd place in the Global Ranking of Defense News Top 100 Defense Industry Companies in 2020.[30]

In 1998, another milestone was the official declaration of the Turkish Defense Industry Policy and Strategy Principles, which covered topics such as primary domestic private sector, international cooperation, and potential export regime. After the 2000s, Türkiye adopted a new phase of defense strategy, one which replaced USA and NATO source-based procurement/import with self-sufficiency-based investment and production.[31] In order to execute this strategy, a compromise was reached by way of the Defense Industry Executive Committee (SSİK) as an additional event in 2004, resulting in the conversion from a threat-based procurement model to a talent-based procurement model based on the maximum use of national resources and domestic production through original design in order to meet TAF's needs. Obtaining the fruits of these action plans began no later than 2010, with the national warship program MİLGEM (covering 65% of the combat management systems produced in Türkiye), the national attack helicopter project ATAK, the national battle tank project ALTAY, the uncrewed aerial vehicle ANKA, and the training aircraft HURKUŞ being among the main achievements of the Turkish defense industry over the years.

Meanwhile, following the 2003 Iraq War and the 2011 post-Arab spring geopolitical turmoil, the changing security landscape and regional disorder forced Türkiye to take essential fundamental precautions in its defense and military strategies. In the process of executing these strategies, improving military innovative competence through domestic and national defense industry capabilities played a significant role in the trend-setting transformation process of the Turkish defense industry. Furthermore, by boosting its political influence in the regional and international arena, Türkiye's claim of being a confident regional player also stimulated this remarkable rise in the Turkish defense industry architecture. In the end, the evolution of the more powerful national defense industry sector has been integral to the Turkish defense strategy. The Turkish defense industry started to shape the priorities of its own military requirements within the framework of these domestic and international developments and new strategic vision.

Realizing the necessity and strong impact of the competitive defense industry after the 2010s, Türkiye focused on national defense technology research and development to overcome its own strategic defense issues. But in reality, foreign market dependency was still inevitable despite the ongoing defense industry integration with the domestic industry and increasing domestic-input joint projects. For instance, although the percentage of defense needs met from domestic sources had increased from 41.6% in 2007 to 54% in 2011, foreign market dependency was still a big concern when compared to the respective accrued rates of 85% and 95% for developed countries.[32]

Within the framework of defense systems' complex, large-scale, and software-intensive nature, another critical factor for increased competitiveness is the well-known fact of clustering actors within the ecosystem in a specific geographic area. Therefore, Türkiye has begun to establish defense industry firms and supporting industries under the cluster of names shown in Table 2.3 to increase the capability of the national defense industry.

Naturally, the main drivers of these sustainable and competitive national defense industry companies are qualified human and intellectual capital/resources. According to SASAD yearly reports, the following figures in Table 2.4 also indicate how the Turkish national defense industry has evolved.

The figures for rate of change, turnover, and employment stand out in the performance of the sector and evolved over the years. The COVID-19 pandemic also affected the 2019 turnover per capita compared to 2020 data with a decrease of 29.22%, unlike sector employment which had increased by 5.14%.

Table 2.3 Defense Industry Clusters in Türkiye[1]

Name	Defense, Aviation, Space Clustering Association (SAHA Istanbul)	BTSO Space and Aviation Cluster	Eskişehir Aviation Cluster (ESAC)	Teknokent Defense Industry Cluster (TSSK)	Aerospace Clustering Association (ACA)	OSTIM Defense and Aviation (OSSA)
Specialty	Defense, aviation ard space	Space and aviation	Aviation	Defense and security	Aviation and space	Defense and aviation
Location	İstanbul	Bursa	Eskişehir	Ankara	İzmir	Ankara
Foundation Year	2015	2013	2011	2010	2010	2008
# of Companies	603	75	31	135 members Σ 330+ companies	78 members (11,000+personnel)	191 (10,000+personnel)

Created by the author. Sources: www.sahaistanbul.org.tr; www.basdec.org; http://tssk.odtuteknokent.com.tr; www.eacp-aero.eu/members. html; https://www.eosb.org.tr/kumelenmeler/esac_eskisehir_aviation_cluster_eskisehir_havacilik_kumelenmesi_dernegi_50.html.

1 Compiled from each cluster's internet sites. SAHA Istanbul (Defence, Aviation and Space Clustering Association) https://www.sahaistanbul.org.tr; The Bursa Space Aviation Defense Cluster (BASDEC), http://www.basdec.org; Teknokent Defence Industry Cluster, http://tssk.odtuteknokent.com.tr; Ostim Defense and Aviation Cluster, https://www.ostimsavunma.org; The European Aerospace Cluster Partnership (EACP) https://www.eacp-aero.eu/members.html; ESAC (Eskişehir Aviation Cluster), https://www.eosb.org.tr/kumelenmeler/esac_eskisehir_aviation_cluster_eskisehir_havacilik_kumelenmesi_dernegi_50.html. accessed on 29.09.2021.

Table 2.4 Turkish Defense Industry Employment Figures[1]

Year	Total Employment	Rate of Change	Turnover Per capita ($)	Change Rate
1998	31,591	-	-	-
2019	73,771	233%	147,539	-
2020	77,566	5%	114,171	-29.22%

Created by the author. Sources: www.tobb.org.tr/Documents/yayinlar/Savunma.pdf, https://www. sasad.org.tr/en/sasad-defense-and-aviation-industry-performance-report.

1 Compiled from TOBB, Türk Savunma Sanayii Sektör Raporu (2007), https://www.tobb.org. tr/Documents/yayinlar/Savunma.pdf, pp. 15–16 and SASAD (Defense and Aerospace Industry Manufacturers Association), Defence Aerospace Sector Performance 2020 Report, https://www.sasad.org.tr/en/sasad-defense-and-aviation-industry-performance-report, accessed on 29.09.2021.

Parallel to these developments, Türkiye has no more time to waste for implementing the necessary strategies. Türkiye's Strategic Vision 2023 project declared the defense industry as the leading sector. The aim is to have all defense-based demands supplied by the domestic market through national technologies. Accordingly, the 10th (2014–2018) and 11th National Development Plans (2019–2023) seized upon the understanding of continuous development for meeting domestic defense needs, ensuring use of all possible opportunities to the maximum extent. The other main objective of these plans is to maximize system-, subsystem-, and component-level defense exports by strengthening the defense industry ecosystem and disseminating skills to the civilian sector. Moreover, the Technology Acquisition Roadmap was also created to systematically monitor technological progress by the Presidency of Strategy and Budget.

In this rapidly changing strategic environment, the defense industry has been declared and modeled as an integral part of Türkiye's hard power instruments, able to benefit from more freedom of strategic maneuverability against security threats in the region, especially since 2016, due to the civil war in Syria itself being a significant threat to Türkiye's national security, including PKK/YPG and DAESH terrorist activities in northern Iraq. DAESH's increasing number of activities directly targeting Türkiye in particular, as well as the lack any tangible support from NATO against these threats have forced Türkiye to take necessary precautions. In order to neutralize these terrorist organizations and restabilize the national security environment, Türkiye has had to conduct successive and assertive military cross-border operations in Syria and Iraq as a final option. As such, Operation Euphrates Shield (August 2016–March 2017), Operation Olive Branch (January–March 2018), Operation Peace Spring (October–November 2019), and Operation Spring Shield (February–March 2020) were the field results of the successful foreign policymaking of Türkiye's newly accomplished self-sufficient military and defense industry strategy. The counter-terrorism efforts that evolved into warfare have also demonstrated how vitally important the principle of indigenousness is.

The combination of a strengthened defense industry and military activism have paved the way for an assertive posture in the defense and military domain, one that has ultimately changed Türkiye's military and defense strategy.[33] The eventual deterioration of Türkiye's diplomatic relations with supplier countries in the defense industry and the rise of tactical and strategic divergences regarding the Syrian war and other regional crises have had a trigger effect on Türkiye's military and defense strategy as Turkish decision-makers lost their trust in the value of partnership with Western countries.[34]

Recent Development Insights

The increased tensions in international relations have caused a mutual distrust between Türkiye and other allied countries. Operations in northern Syria and primarily the purchase of the Russian-made S-400 surface-to-air missile system have resulted in sanctions on Türkiye's Presidency of Defense Industries institution. US and European partners are no longer reliable for sustaining Türkiye's required military equipment. One of the results of these implemented sanctions was the US excluding Türkiye from the F-35 program despite being an original contributor and manufacturer. Another result was Germany's reluctance to modernize Türkiye's Leopard 2 tanks inventory. These incidents have reinforced Türkiye's need to become more independent in developing its defense industry and its own defense capabilities. Türkiye has been replacing the AH-1 attack helicopters with Turkish Aerospace Industries T129 ATAK and planning to put its first domestically produced Altay tanks in the field at the beginning of 2023, in addition to upgrading its Leopard 2s without German assistance.

Concerning navy projects, Türkiye has launched the first of four planned İstanbul-class frigates and TCG Anadolu amphibious assault ships to function as light aircraft carriers with the ability to launch new Turkish-made aircraft, this in addition to four Ada-class corvettes. Furthermore, another project for an uncrewed surface vessel armed with six guided missiles named ULAK having started its sea trials is worth mentioning as a new design for sea platforms. Moreover, Türkiye is constructing two modified Ada-class corvettes for Pakistan's navy (with Pakistan building two more itself under license) and at least one Ada-class corvette for Ukraine. Türkiye also has plans for a mobile naval mine for use on its amphibious assault ships that can be used for surveillance and to attack ships and uncrewed fighter jets, as well as strike aircraft; officials say it will be able to carry 30 to 50 drones.[35]

Regarding UAVs, the Turkish-made Bayraktar TB2 uncrewed combat aerial vehicles are Türkiye's most impressive initiative in becoming a leader in uncrewed systems, besides several ongoing uncrewed systems in development. These systems were real force multipliers to Azerbaijan's victory in the recent war with Armenia. To suggest that other countries looking for an aerial advantage to solve their security problems are also seeking ways to own TB2 UAVs would not be wrong. Additionally, Türkiye has begun mass production of a larger uncrewed combat aircraft with a payload of 1.5 tons known as Akıncı.

There has been a heavy emphasis on the idea that Türkiye was going to develop its indigenous capabilities so as to be able to become both more effective as a military producer and less dependent on foreign sources, as well as to gain the potential for export-driven growth. As a result, Türkiye is determined to grow its domestic defense industry. The aforementioned increase in domestic defense output has allowed Türkiye to become a flourishing weapons exporter within the scope of its having built 28 different country defense industry markets. The global arms trade tracker SIPRI recently listed Türkiye as one of the three fastest-growing arms exporters, with the volume of its exports having increased by 86% in the latter half of the 2010s. Over the same period, it rose six ranks to become the world's 13th largest arms exporter, while going from the 6th largest arms importer to the 20th; these are promising developments for the future of the Turkish defense industry.

Feeling spurned by its traditional partners, Türkiye seeks partnering countries like South Korea and Ukraine to boost its remaining technological gaps. However, it should be mentioned that Türkiye still has some Western technology-based dependence and currently is not yet self-sufficient. For example, Türkiye still uses US-made engines in the T129 ATAK, German guns for the Altay tank, and German air-independent propulsion systems for the new Reis-class submarines. Moreover, concerning Türkiye's exclusion from the F-35 project, even if Türkiye were to strive to build her own fifth-generation TF-X (Turkish Fighter X) to replace the F-35, the reality of the F-16 life-extension program suggests this might not be a possibility. These issues have two different consequences. On one hand, they have fueled Turkish ambition to build a more independent defense industry. On the other hand, they have triggered a debate about whether Türkiye has changed its overall strategic stance. However, the time is still too early to reach a proper conclusion, and claiming that Türkiye is changing its geostrategic direction may also be an exaggerated assessment.

Conclusion

Previously, Türkiye had adopted and practiced her defense and security policies, structured upon deterrability, with conventional weapons systems mainly produced by NATO allies. However, beginning in the 2000s and in parallel with rapidly changing environmental and technological global advances, this trend of defense industry principles has been realigned and updated toward adopting domestic production, national self-sufficiency, and competency. This development has resulted from the sanctions, embargoes, and technical problems that put the country in a vulnerable position with regard to separatist terrorism and other regional threats. As such, coproduction, cooperation/joint works, technology transfer, and removal of sole-source supply dependence/foreign purchasing became essential themes in the new generation defense industry strategy to focus on managing and enhancing strategic autonomy. In summary, Türkiye has been seeking and striving to be less dependent on outsiders in the defense industry. This strategy requires paving the way for building a sustainable defense industry by investing and expanding the production of new ships, tanks, and missiles to bolster and outfit its military gear and capabilities.

Türkiye has constantly emphasized possessing effective, deterrable armed forces with advanced operational capabilities for maintaining and increasing its security. One of the capabilities that can provide this security is to continuously develop the domestic defense industry and infrastructure to reduce foreign dependency, in line with domestic and national opportunities, by focusing on R&D investments so as to equip the Turkish Armed Forces with modern technology. To conclude that the Turkish defense industry is still in the process of its evolution to meet all kinds of needs, not only for the Turkish Armed Forces but also for the NATO alliance and other potential customer states, would not be wrong. However, this progress is only possible and achievable by both being a decent operator and a unique producer of military innovation capabilities.

Notes

1 Chris Hedge, *What Every Person Should Know About War?* Free Press, New York, 2003, p. 1.
2 Stockholm International Peace Research Institute (SIPRI), 2019, https://www.sipri.org/databases/milex, accessed on 29.09.2021.
3 SIPRI, 1988–2019, https://www.sipri.org/media/press-release/2020/global-military-expenditure-sees-largest-annual-increase-decade-says-sipri-reaching-1917-billion, accessed on 29.09.2021.
4 SIPRI 2021, https://www.sipri.org/media/press-release/2021/world-military-spending-rises-almost-2-trillion-2020, accessed on 29.09.2021.
5 Erman Erbaykal, *Türkiye'de Savunma Harcamaları ve Ekonomik büyüme İlişkisi*, (Master of Arts thesis), Balıkesir Üniversitesi, 2007, p. 13.
6 Özlem Durgun ve Mustafa Caner Timur, "Savunma Harcamaları ve Ekonomik Büyüme İlişkisi: Türkiye Analizi", *Dumlupınar Üniversitesi Sosyal Bilimler Dergisi*, 54, 2017, p. 127.
7 M. Hakan Özbaran, "Türkiye'de Kamu Harcamalarının Son Beş Yılının Harcama Türlerine Göre İncelenmesi", *Sayıştay Dergisi*, 53, 2004, p. 80.
8 Michael Brzoska, *Handbook of Defense Economics I*, Elsevier, Amsterdam, 1995, p. 49.
9 Filiz Giray, "Savunma Harcamaları ve Ekonomik Büyüme", *Cumhuriyet Üniversitesi İktisadi ve İdari Bilimler Dergisi*, 5:1, 2004, p. 188.
10 Saadet Değer and Somnath Sen, "Military Expenditure, Spin-off and Economic Development", *Journal of Development Economics*, 13:1–2, 1983, pp. 69–70.
11 Nezihi Çakar, "A Strategic Overview of Turkey", *Journal of International Affairs*, III.2, 1998, p. 1.
12 SIPRI, 2020, https://www.sipri.org/databases/milex, accessed on 29.09.2021.
13 Defence Expenditure of NATO Countries, 2019, https://www.nato.int/cps/en/natohq/news_171356.htm, accessed on 29.09.2021.
14 J. Paul Dunne and Mehmet Uye, "Military Spending and Development," Working Papers 0902, Bristol Business School, University of the West of England, Bristol, 2009, p. 2.
15 SIPRI *Military Expenditure Database*, https://www.sipri.org/sites/default/files/SIPRI-Milex-data-1949-2018_0.xlsx, accessed on 29.09.2021.

16 Aziz Akgül, *Dünyada Savunma Harcamaları ve Savunma Sanayilerinin Yapısı*, Özel Yayın, Ankara, 1987, p. 5.

17 *Stockholm International Peace Research Institute Top 100 Arms-Producing and Military Services Companies*, 2019, https://www.sipri.org/sites/default/files/2019-12/1912_fs_top_100_2018.pdf, accessed on 29.09.2021.

18 Tolga Öz, *Reverse Logistics and Applications in the Defense Industry*, *(Master Thesis)*, Dokuz Eylül Üniversitesi, İzmir, 2007, p. 112.

19 *Defense News Top 100 for 2019*, 2019, https://people.defensenews.com/top-100/, accessed on 29.09.2021.

20 M.Ali Birand, *Emret Komutanım*, Milliyet Yayınları, 8.Baskı, İstanbul, 1986, p. 371.

21 Elliot Hen-Tov, "The Political Economy of Turkish Military Modernization", *MERIA Journal*, 08:04, 2004, p. 4.

22 Anthony Forster, *Armed Forces and Society in Europe*, Palgrave Macmillan, New York, 2006, p. 156.

23 History, Art & Archives United States House of Representatives, "Historical Highlights: The Lend-Lease Act of 1941" (March 11, 1941), http://history.house.gov/Historical-Highlights/1901-1950/The-Lend-Lease-Act-of-1941/, accessed on 29.09.2021.

24 Çağrı Erhan, "Turkey's relations with US and NATO", in (ed.), Baskın Oran, *Turkish Foreign Policy, Facts, Documents and Remarks from Turkish War of Independence till Nowadays Vol I*, Issue 6, İletişim Yayınları, İstanbul, 2002, pp. 540–541. Hüseyin Bağcı, *Turkish Foreign Policy in 1950s*, Issue 3, METU Publishing, Ankara, 2007, pp. 5–7, 37–38.

25 "The Defense and Economic Cooperation Agreement-U.S. Interests and Turkish Needs", Report by the Comptroller General of the United States, May 7, 1982 https://www.gao.gov/assets/140/137457.pdf, accessed on 29.09.2021.

26 Joshua Kucera, *U.S. Denies Turkey Leftover Warships*, Jan 5, 2015. https://eurasianet.org/us-denies-turkey-leftover-warships, accessed on 21.10.2021.

27 Turkish Armed Forces Foundation, https://www.tskgv.org.tr/en/about-us/history, accessed on 29.09.2021.

28 Presidency of The Republic of Türkiye Presidency of Defence Industries, https://www.ssb.gov.tr/Website/contentList.aspx?PageID=39&LangID=2, accessed on 29.09.2021.

29 Turkish Defense and Aerospace Industry Manufacturers Industry, https://www.sasad.org.tr/en/about-us, accessed on 29.09.2021.

30 Defense Technologies and Engineering Inc., https://www.stm.com.tr/en/who-we-are/about-us, accessed on 29.09.2021.

31 Hüseyin Bağcı and Çağlar Kurç, "Turkey's strategic choice: buy or make weapons?", *Defense Studies*, 17:1, December 2016, pp. 38–62; Arda Mevlütoğlu, "Commentary on Assessing the Turkish Defense Industry: Structural Issues and Major Challenges", *Defense Studies*, 17:3, July 2017, pp. 282–294.

32 Onuncu Kalkınma Planı 2014–2018, https://www.sbb.gov.tr/wp-content/uploads/2021/12/Onuncu_Kalkinma_Plani-2014-2018.pdf [accessed on 29 September, 2021].

33 Arda Mevlütoğlu, "Türkiye'nin Savunma Reformu: Tespit ve Öneriler", *SETA Analiz*, No. 164 (August 2016), retrieved from https://setav.org/assets/uploads/2016/09/20160901200637_turkiyenin-savunma-reformu-pdf1.pdf., accessed on 29.09.2021.

34 "Turkish Defence Industry Summit Held on Presidential Complex," *Turkey Defense*, (December 12, 2018), retrieved from https://www.defenceturkey.com/en/content/turkish-defence-industry-summit-held-on-presidential-complex-3329, accessed on 29.09.2021.

35 *Business Insider*, https://www.businessinsider.com/turkey-is-modernizing-its-military-to-send-message-to-nato-2021-8, accessed on 29.09.2021.

Bibliography

Akgül, Aziz, *Dünyada Savunma Harcamaları ve Savunma Sanayilerinin Yapısı*, Özel Yayın, Ankara, 1987.

Ateş, Barış, *Soğuk Savaş Sonrası Dönemde Askeri Değişim: NATO Orduları ve Türk Silahlı Kuvvetleri Üzerine Karşılaştırmalı Bir Analiz*, Unpublished doctoral dissertation, Gazi University, Ankara, 2014.

Bağcı, Hüseyin and Kurç, Çağlar, "Turkey's Strategic Choice: Buy or Make Weapons?" *Defense Studies*, 17:1 December 2016, pp. 38–62.

Bağcı, Hüseyin, *Turkish Foreign Policy in 1950s*, Issue 3, METU Publishing, Ankara, 2007.

Birand, M. Ali, *Emret Komutanım*, Milliyet Yayınları, 8.Baskı, İstanbul, 1986.

Brzoska, Michael, *Handbook of Defense Economics I*, Amsterdam: Elsevier Science B.V., 1995.

Business Insider, https://www.businessinsider.com/turkey-is-modernizing-its-military-to-send-message-to-nato-2021-8.

Çakar, Nezihi, "A Strategic Overview of Turkey", *Journal of International Affairs*, III:2, 1998, pp. 1–5.

Defence Expenditure of NATO Countries, 2020, https://www.nato.int/cps/en/natohq/news_171356.htm.

Defense News Top 100 for 2019, 2019, https://people.defensenews.com/top-100/.

Defense Technologies and Engineering Inc., https://www.stm.com.tr/en/who-we-are/about-us.

Değer, Saadet and Sen, Somnath, "Military Expenditure, Spin-off and Economic Development", *Journal of Development Economics*, 13:1–2, 1983, pp. 69–70, https://doi.org/10.1016/0304-3878(83)90050-0

Dunne, J. Paul and Üye, Mehmet, "Military Spending and Development", *Working Papers 0902*, Bristol Business School, University of the West of England, Bristol, 2009.

Durgun, Özlem and Timur, Mustafa Caner, "Savunma Harcamaları ve Ekonomik Büyüme İlişkisi: Türkiye Analizi", *Dumlupınar Üniversitesi Sosyal Bilimler Dergisi*, 54, 2017, pp. 118–139.

Erbaykal, Erman, *Türkiye'de Savunma Harcamaları ve Ekonomik büyüme İlişkisi*, *(Master Thesis)*, Balıkesir Üniversitesi, Balıkesir, 2007.

Erhan, Çağrı "Turkey's Relations with US and NATO", in (ed.) Baskın Oran, *Turkish Foreign Policy, Facts, Documents and Remarks from Turkish War of Independence till nowadays Vol I*, Issue 6, İletişim Yayınları, İstanbul, 2002, pp. 540–541.

ESAC (Eskişehir Aviation Cluster), 2011. https://www.eosb.org.tr/kumelenmeler/esac_eskisehir_aviation_cluster_eskisehir_havacilik_kumelenmesi_dernegi_50.html.

Forster, Anthony, *Armed Forces and Society in Europe*, Palgrave Macmillan, New York, 2006.

Giray, Filiz, "Savuma Harcamaları ve Ekonomik Büyüme", *Cumhuriyet Üniversitesi İktisadi ve İdari Bilimler Dergisi*, 5:1, 2004, pp. 181–199.

Hedge, Chris, *What Every Person Should Know About War?* Free Press, New York, 2003.

Hen-Tov, Elliot, "The Political Economy of Turkish Military Modernization", *MERIA Journal*, 08:04, 2004, pp. 1–28.

History, Art & Archives United States House of Representatives, "Historical Highlights: The Lend-Lease Act of 1941" (March 11, 1941), http://history.house.gov/Historical-Highlights/1901-1950/The-Lend-Lease-Act-of-1941/.

IISS (The International Institute for Strategic Studies), The Military Balance 2021, Routledge, 2021. https://doi.org/10.4324/9781003177777

IISS (The International Institute for Strategic Studies), The Military Balance 1991–1992, Routledge, 1991.

Kucera, Joshua "US Denies Turkey Leftover Warships", January 5, 2015. https://eurasianet.org/us-denies-turkey-leftover-warship.

Mevlütoglu, Arda, "Commentary on Assessing the Turkish Defense Industry: Structural Issues and Major Challenges", *Defense Studies*, 17:3, July 2017, pp. 282–294.

Mevlütoğlu, Arda, "Türkiye'nin Savunma Reformu: Tespit ve Öneriler", *SETA Analiz*, No. 164 (August 2016), retrieved from https://setav.org/assets/uploads/2016/09/20160901200637_turkiyenin-savunma-reformu-pdf1.pdf

Onuncu Kalkınma Planı 2014-2018, 2013, https://www.sbb.gov.tr/wp-content/uploads/2021/12/Onuncu_Kalkinma_Plani-2014-2018.pdf

Ostim Defense and Aviation Cluster, 2008, https://www.ostimsavunma.org.

Öz, Tolga, *Reverse Logistics and Applications in the Defense Industry*, *(Master Thesis)*, İzmir, Dokuz Eylül Üniversitesi, 2007.

Özbaran, M. Hakan, "Türkiye'de Kamu Harcamalarının Son Beş Yılının Harcama Türlerine Göre İncelenmesi", *Sayıştay Dergisi*, 53, 2004, pp. 75–111.

Presidency of The Republic of Türkiye Presidency of Defence Industries, 2018, https://www.ssb.gov.tr/Website/contentList.aspx?PageID=39&LangID=2.

SAHA Istanbul (Defence, Aviation and Space Clustering Association), 2015, https://www.sahaistanbul.org.tr.

SASAD (Defense and Aerospace Industry Manufacturers Association), Defence Aerospace Sector Performance 2020 Report, https://www.sasad.org.tr/en/sasad-defense-and-aviation-industry-performance-report.

SIPRI (Stockholm International Peace Research Institute), *Top 100 Arms-Producing and Military Services Companies*, 2019a, https://www.sipri.org/sites/default/files/2019-12/1912_fs_top_100_2018.pdf.

SIPRI (Stockholm International Peace Research Institute), 2019b, https://www.sipri.org/databases/milex.

SIPRI (Stockholm International Peace Research Institute), 2020, https://www.sipri.org/databases/milex.

SIPRI (Stockholm International Peace Research Institute), 1988–2019, https://www.sipri.org/media/press-release/2020/global-military-expenditure-sees-largest-annual-increase-decade-says-sipri-reaching-1917-billion.

SIPRI (Stockholm International Peace Research Institute), 2021, https://www.sipri.org/media/press-release/2021/world-military-spending-rises-almost-2-trillion-2020.

SIPRI (Stockholm International Peace Research Institute), *Military Expenditure Database*, https://www.sipri.org/sites/default/files/SIPRI-Milex-data-1949-2018_0.xlsx.

Teknokent Defence Industry Cluster, http://tssk.odtuteknokent.com.tr.

The Bursa Space Aviation Defense Cluster (BASDEC), http://www.basdec.org.

"The Defense and Economic Cooperation Agreement-U.S. Interests and Turkish Needs", Report by the Comptroller General of the United States, May 7, 1982 https://www.gao.gov/assets/140/137457.pdf.

The European Aerospace Cluster Partnership (EACP) https://www.eacp-aero.eu/members.html

TOBB, Türk Savunma Sanayii Sektör Raporu (2007) https://www.tobb.org.tr/Documents/yayinlar/Savunma.pdf

Turkish Armed Forces Foundation, https://www.tskgv.org.tr/en/about-us/history.

Turkish Defense and Aerospace Industry Manufacturers, https://www.sasad.org.tr/en/about-us.

"Turkish Defence Industry Summit Held on Presidential Complex," *Turkey Defense*, (December 12, 2018), retrieved from https://www.defenceturkey.com/en/content/turkish-defence-industry-summit-held-on-presidential-complex-3329

3 Breaking Down Bureaucratic Politics in the Turkish Aerial Defense Industry

From Hürkuş to Bayraktar

Mehmet Mert Çam

Introduction

What lies at the heart of international relations is not where states position themselves. The actors instead assess a country, either through the lens of psychological characteristics or through the material factor of national power. Türkiye prioritized improvements in the defense industry (i.e., the manufacture, development, and use of UAVs) during the assessment phase carried out by the system players. Having proven their efficiency in conflict regions from nearly half a century of counterterrorism operations with the civil wars in Syria and Libya, the Armenian-occupied Nagorno-Karabakh, and Russian separatists in Ukraine, Turkish UAVs have drawn the attention of foreign media and academia.

Despite the notion in the Turkish defense industry that *no good deed goes unpunished*, which has devolved from the past to the present into a kind of learned helplessness, the story of the Bayraktar UAVs, where a private enterprise embarked on this desperate and dangerous path in the defense industry, clearly also deserves academic attention. Yet under these circumstances, the following question arises, "How did the Bayraktar family succeed in producing a weapon system that could change regional balances despite the negative examples of the past, the monopoly of several countries in the defense industry, and the change-resistant bureaucratic structure?"

This chapter focuses on the process of producing and developing new technology, in contrast to the other chapters which deal with the effects of new military technologies in other areas. However, this focus turns toward the personal experiences and inner worlds of individuals who have experienced the development process from the beginning, rather than concentrating on the technical issues. Therefore, the chapter aims to offer insight into the future by addressing the past.

Bureaucracy and the Case of Hürkuş

Weberian bureaucracies are closed to innovation due to their organizational culture. Adherence to administrative traditions, strict hierarchies, the unquestionable authority of written rules, and lack of initiative due to

DOI: 10.4324/9781003327127-4

uncompromising personalities result in the preservation of the *status quo*. In other words, resistance to innovations through business as usual.[1] Apart from these bureaucracies' general characteristics, their geographical location, culture, and religion are other parameters that significantly affect how they function. The sociological structure and psychological status of the individuals involved in a bureaucracy, who represent society within the organization's culture[2], lead the system to change, for better or worse.

The bureaucratic politics approach is a conceptual framework Graham T. Allison formed for the first time in his article "Conceptual Models and the Cuban Missile Crisis" in 1969. It states that the bargaining-based approach which occurs among the various organizations and individuals within a state's public policy, represents the bureaucracy in one regard, and suggests that political issues originate from those bureaucratic actors who are positioned at low, middle, and high levels.[3] Allison evaluated the bureaucratic politics approach as a contrast to the understandings of realist and rationalist political decision-making, and one in which actors who are involved in the game have various preferences, abilities, and power positions, and therefore choose strategies and political goals based on which outcomes will best serve their corporate and personal interests. Bargaining later turns into a pluralist process of give-and-take, one that reflects the prevailing rules of the game and power relations among the actors. Non-optimal results emerge because this process has no supervision and expert/rational decisions are overlooked during the process. Actors being elected or appointed, having high or low positions, or being experienced or inexperienced in their roles affects their interests and bargaining positions.[4] This situation may cause individual or organizational interests to be prominent at the micro-level, as opposed to *public interest*, as a result of this very intricate process.

Allison's study serves as a lens for the decision-making mechanism of the United States in the process of determining domestic and foreign policy. However, one may infer that this paradigm also generally applies to Türkiye, given the universality of contemporary bureaucracies on a fundamental level and Türkiye's status as a strict NATO and US ally during the Cold War. Apart from the interwar period, Türkiye's strict Western approach to defense policies in addition to the "bureaucratic warfare" between organizations have created non-optimal results.

Foucault noted that restoring something is more difficult than building something from scratch.[5] In fact, certain sectors, in particular the military industry, could have taken steps to quickly shut down the power imbalance that emerged after the Ottoman Empire's fall and during the early Republican era. However, during that time when a new bureaucracy had replaced the old (which, within a decade, then prevailed again), the old order was able to find its own sphere of influence through different paradigms. In this regard, what Vecihi Hürkuş, one of the first Turkish pilots, experienced as an engineer and entrepreneur is noteworthy. Before mentioning this, however, one must be familiar with Vecihi Hürkuş and the motivation behind his contributions to the Turkish air defense industry.

Vecihi Hürkuş made his first combat flight on the Caucasian Front in World War I, during which he shot down a Russian aircraft on September 26, 1917. This incident was the first air victory in Turkish war history, and afterward he was captured by the Russians and deported to Nargin Island in the Caspian Sea. After four months, he managed to escape and returned to Türkiye by way of Iran.[6] When he was in charge of the İstanbul Air Defense Company, he designed an attack aircraft "both due to the imperative for the defense of the homeland and its usefulness for developing national aviation".[7] The fact that he had witnessed this need during the air raids he had launched against the Russians on the Caucasian Front is important in terms of *on-site observation*. However, in wartime conditions this idea was considered ridiculous and utopic by those around him as well as by his superiors. Based on his experience at the front at the time, he had also noticed how the aircraft's front-mounted machine guns damaged the propeller. Despite reporting this manufacturing fault to his superiors with a proposed solution, they remained unresponsive. In his memoirs, Hürkuş stated that they had responded, "If this were a malfunction, the manufacturer would have anticipated and fixed it accordingly"[8] and remained indifferent, despite the proof of his claim. This is a typical example of the bureaucratic approach.

Vecihi Hürkuş, on his own, was able to repair and fly two Nieuport-type attack aircraft seized from the Russians near the end of the war. He also conducted reconnaissance and observation activities in each battle using resources obtained from the enemy during the War of Independence and produced wing repair fabric[9] for captured aircraft. During the Great Offensive, Hürkuş was able to first enter into İzmir airspace and managed to shoot down a Greek aircraft.[10] Vecihi Hürkuş then took advantage of the power gap existing between the War of Independence and the proclamation of the Republic to start designing the Vecihi K-VI aircraft with the support and encouragement of Colonel Muzaffer (Ergüder), the inspector of the air force at that time. He brought a nine-seat passenger plane that the Italians had abandoned in Edirne to İzmir and quickly completed the plane's assembly using engines from seized Greek planes. A technical committee was established to obtain a flight certificate for the Vecihi K-VI, the first Turkish aircraft whose fuselage, wings, and other parts were manufactured using domestic materials. However, no flight permit could be obtained due to the lack of staff technically qualified to provide the flight permit. After the head of the committee suggested, "We cannot give this permission to you. If you trust your aircraft, go fly and relieve us as well," he made his 15-minute flight on January 28, 1925. The aircraft reached a speed of 207 kmh, which was faster than the Avions Caudron C-59 aircraft that Türkiye had bought from France at the time.[11]

Vecihi Hürkuş's success did not go unpunished, as Colonel Muzaffer, from whom he had received support for the project, confiscated his aircraft for flying it without permission. As a result, he had to submit his resignation from the Air Force and never got his Vecihi K-VI aircraft back. Nevertheless,

Vecihi Hürkuş carried out his activities in line with the goal of raising an aviation generation to contribute to establishing a national air industry. He was one of the five members of the *Türk Tayyare Cemiyeti* [Turkish Aeroplane Society] and started working as a highly experienced pilot at TOMTAŞ (Turkish Aircraft and Motor Incorporated Company), a Turkish-German partnership that was established in the spring of 1925. In Hürkuş's following words, the presence of TOMTAŞ and the supply of aircraft to the Turkish Air Force had disturbed domestic and foreign circles in terms of competition, and interest groups did not want a national organization to exist:

> The two well-known companies Yılmazlar and Cudiler almost every day shuttled back and forth in the bureaus of the Undersecretariat for Air of the Ministry of National Defense and in particular did not hesitate to influence the personnel of the technical bureau. They knew that, once our air force possessed aircraft made in Türkiye, there would be no need to buy foreign aircraft.[12]

According to this statement, even though the maintenance and repair of the German-originating JU A-20s in the inventory was the responsibility of TOMTAŞ as per the Ministry of National Defense, this contract was cancelled after a while through pressure from representatives of foreign companies, with TOMTAŞ going bankrupt and shutting down in 1928. Hürkuş described the establishment, saying, "That which has gnawed the Turkish Aviation from the inside since day one is the bureaucracy and interest groups themselves".[13] When considering that 80% of today's aircraft factories worldwide had opened after 1940, the domestic production-oriented moves made in the first quarter of the 20th century, which is a very early period even for world aviation, are remarkable. The presence of technically ignorant and opportunist deputies of TOMTAŞ's management, the ambition for financial gain in foreign companies and their partners in Türkiye, and the jealousy from others' successes in the lower and middle branches of the bureaucracy that lacked patriotic consciousness had caused the project to come to a standstill, despite Mustafa Kemal Atatürk's support.[14]

The Turkish Aeroplane Society did not allow Vecihi Hürkuş to design his Vecihi XIV, a sport-training aircraft; however, he was able to manage the plane's completion in a workshop on the upper floor of a store in Kadıköy, İstanbul between June 19 and September 16, 1930. Two days later, Hürkuş went to Ankara with his own plane and received great attention and appreciation from everyone, particularly the top brass of the state officials. After that, however, he hit against the wall of bureaucracy once again with the response "It cannot be certified as airworthy because of the lack of technical personnel and equipment for determining [the aircraft's] aerodynamic qualities"[15] he received from the Ministry of National Defense, Undersecretariat for Air to whom he had applied for a flight certificate. Nonetheless, he did not give up and went to Czechoslovakia with the support of Marshal Fevzi

Çakmak, the Chief of the Defense Staff at the time. There he managed to obtain a flight certificate from the International Commission for Air Navigation (CINA).

Upon his return to Türkiye, he started promotional trips around Anatolia with his Vecihi XIV to raise public awareness about aviation, gave lectures at schools to create an "aviation generation" in İstanbul, and even founded the Vecihi Civil Aviation School in Kadıköy in 1932. Vecihi Hürkuş provided free education to 12 students, including Bedriye Gökmen[16] as the first Turkish airwoman and Mrs. Eribe who was the first Turkish airwoman casualty, and founded the Vecihi-Faham Aircraft Construction Factory.

The most exciting invention from Vecihi Hürkuş, who designed and manufactured four more aircraft in two years, was an aquaplane. He tried to sell this aquaplane, called "Hovercraft", to the Turkish naval forces; however, neither the state nor the private sector showed interest, which was similar to what he had experienced with his previous aircraft. Hürkuş ensured the sustainability of the education at his school with donations raised from various advertisements and philanthropic businessmen. Despite the factory's failure to receive orders, with the support of businessman-politician Nuri Demirağ it built Türkiye's first closed-cabin passenger plane, the Vecihi-XVI Nuri Bey.[17] However, due to financial difficulties, the school and the factory were closed in 1935.

President Mustafa Kemal Atatürk's initiation of the Türkkuşu project in the same year, with the aim of improving civil aviation in Türkiye, and his desire to have this project benefit from Vecihi Hürkuş led Hürkuş to return to the *Türk Hava Kurumu* [Turkish Air Association]. He was sent to Germany to study engineering again under the instruction of Gazi Pasha in 1937. A year and a half later in February 1939, he received his diploma in aircraft mechanical engineering from the Weimar Engineering School.

Vecihi Hürkuş now believed that no obstacle existed any more to prevent him from building an aircraft; yet he encountered a very different environment upon returning to Türkiye. After the death of Mustafa Kemal Atatürk, the cadres (along with the management style and mentality of the Turkish Air Association) had changed. As a result, Vecihi Hürkuş was exiled to Van by the new order in his organization and was refused an aircraft engineering license with the justification that "one cannot become an engineer in two years"; in fact, he said he was not even allowed to hammer nails.[18]

After World War II, despite his lack of political involvement, he joined the Democratic Party to "destroy the wall of bureaucracy and create wings", but the dominating bureaucracy that prevailed over politics once again triumphed. The impact of the Cold War was significant in this regard. What confirms this argument is the fact that General Zeki Doğan, the then general of the air force who was an army officer at heart, explained his reason for not ordering aircraft from the Turkish Air Association, saying, "If I ordered from your aircraft factory, paying money instead of getting free aircraft from American aid, these people would hang me".[19] The influence of the American Grand Strategy in the region, built on the premise of containing the Soviets during the Cold War, had grown even more since 1952. US National Security Council Document 162/2 clearly states that the United States should "support Türkiye in the political and military fields and assist in the economic and technical fields".[20]

Vecihi Hürkuş established the *Kanatlılar* [Winged] Association in 1947, with the majority of its members being university students, and proceeded to inculcate a passion for flying using an aircraft he had acquired from the Turkish Aeronautical Association. To earn his livelihood and avoid leaving the skies, he flew for different advertising businesses using aircraft he had purchased from the United Kingdom. In 1954, he founded Hürkuş Airlines, purchasing aircraft that Turkish Airlines had put up for sale. Although the contract originally stated unequivocally that the aircraft were ready to fly, the contract turned into "no flight certificate can be granted" as a result of overnight lobbying at the Ministry of Transport. As a result, Hürkuş Airlines, which had been able to start its flights along the İstanbul-Bursa, İzmir-Aydın, and İzmir-Milas routes on April 1, 1955, now faced the General Directorate of State Airlines' wrath. Its flights were first restricted and then blocked, and its planes were left in the open to suffer body damage.

The sabotage and "assassination" attempt against the TC-PER plane, that was under the control of Vecihi Hürkuş and to be transferred from Bursa to Yeşilköy on July 24, 1955, was the final straw. The aircraft had been left in front of the administrative building of the General Directorate of State Airlines, and its left engine broke down immediately after takeoff, forcing Vecihi Hürkuş to land. When the aircraft's gasoline filter was removed, pieces of cotton were found inside, and the safety wires had been removed. However, the perpetrators became lost in the corridors of bureaucracy, despite inquiries and investigations.[21] The aircraft belonging to Vecihi Hürkuş, whose workshop was confiscated after the 1960 Military Intervention, were smashed to pieces, and Hürkuş Airlines vanished. Tirelessly, he then started what he called a dangerous job with his last aircraft and made reconnaissance flights searching for radioactive mines (e.g., thorium, uranium, phosphorous) in Türkiye with permission from the General Directorate of Mineral Research and Exploration. As a "reward" for this service, his commercial pilot's license was canceled in 1966 due to "endangering the safety of people and property on low altitude flights".[22]

Even after the death of Vecihi Hürkuş, the first and only Turkish pilot to have spent 30,000 hours in the air with 102 different types of aircraft, efforts to discredit him continued in various areas. For example, a 1977 Turkish movie portrayed Vecihi as an irrational dreamer who flew over the city aimlessly, causing damage to buildings and people. Even today, this movie is shown on TV channels to deter anyone from daring what Vecihi Hürkuş had attempted.[23]

Türkiye's UAV Adventure and the Case of Bayraktar

Along with Britain and Israel, the "staunch allies" of the United States during the Cold War, Türkiye adhered to the American Grand Strategy under collective security throughout the Cold War. As shown by American intelligence records, the Johnson Letter's success in preventing Türkiye's involvement in Cyprus as a guarantor state established a significant Soviet influence in the eastern Mediterranean at the time.[24] Notably, the letter stated that American equipment could not be used in the likelihood of Turkish interference; if it were, NATO would be unable to assist Türkiye in the event of a Soviet attack in compliance with Article 5.[25] Despite the implicit support of US Secretary of State, Kissinger, in the intervening decade after the Johnson Letter[26] the arms embargo enacted on Türkiye and the problems experienced with military equipment during the operation necessitated the re-establishment of the national defense industry.

The end of the Cold War and the collapse of the Soviet Empire placed Türkiye, which had traditionally purchased weaponry and military equipment from the Western world, particularly the United States, in a difficult position for the second time. New arms dealers emerged for Türkiye, which was fighting against the terrorist organization PKK/KCK[27]; 17 Mi-17 helicopters had even been purchased for the Turkish gendarmerie from Russia, who had been the main enemy in the Cold War. However, these helicopters were unable to be used effectively due to the difficulties experienced during modernization in Russia[28] and the structuring of the Turkish Army in the conception of NATO.[29] During the post-Cold War period, the Turkish defense industry's closest partner was Israel, who had been contracted to upgrade the aircraft, tanks, and even ships for the Turkish Army. Türkiye and Israel conducted joint military exercises and trainings, with the two countries signing the Defense Industry Cooperation Agreement in 1996. According to the agreement, the two countries agreed on issues such as joint action in the defense industry, mutual technology transfer, and joint arms production.[30] In this regard, the two states' common interests have been influential dating back to the Cold War. In the power vacuum created by the establishment of the new world order, both sides acted pragmatically, with Türkiye and Israel forming an alliance against Greece, South Cyprus, and Egypt in the eastern Mediterranean.[31]

In 2001, Türkiye signed an agreement to buy UAVs from Israel. They agreed on ten Herons in exchange for 183 million US dollars in 2005, with the first batch of Herons being delivered by the Israel Aviation and Space Industry (IAI) in 2007.[32] However, the Israeli-made UAVs were a new generation system for intelligence and observation and did not meet expectations. The Herons had begun being used with no clear national doctrine, and three crashed after about a year. Meanwhile, the others either could not take off due to technical malfunctions or were lost while in flight. The fact that Herons that disappeared in the air were found only through signals sent from Israel[33] led to an intelligence vulnerability for Turks. In addition, the inability in the first place to integrate the 120 kg ASELSAN-produced Aselflir thermal imaging system into the Herons had also caused a strategic problem. Despite being included in the contract, Herons' endurance in the air, which required flying at an altitude of 30,000 and flying for 24 hours, had been shortened:

They were only able to reach a maximum altitude of 24,000 feet.[34] In summary, while the Herons were successful in some operations, they were unable to deliver the agreed-upon results.

While negotiations were underway regarding the Herons and other defense problems in Türkiye, a Turkish engineer who had worked as a research assistant at the Massachusetts Institute of Technology (MIT) at that time made his autonomous formation flight with UAVs for the first time at Fort Benning Infantry Base in the early 2000s. As will be discussed in detail below, Selçuk Bayraktar's struggle as the technical director of BAYKAR Tech is very similar to that of Vecihi Hürkuş. While the global order, circumstances, players, and technology had changed over time, the dominant bureaucracy had maintained its presence.

UAVs started to be used by Israel in the last quarter of the 20th century, being in continuous conflict, and by the United States, in particular, primarily to obtain observation reports and intelligence. Fighting against terrorism, Türkiye expressed its interest in this field by purchasing externally, unlike the other two countries. For example, Türkiye bought four GNAT 750 uncrewed scout vehicles in the 1990s.[35] However, just as had happened with the Herons, these vehicles, purchased abroad, were incapable of being used effectively enough. As Selçuk Bayraktar stated in our interview, Türkiye was interested in UAVs in order to turn the equation in its favor, to create asymmetry in an unconventional operation, and to break the other side's will to fight.[36]

Before mentioning the bureaucratic obstacles and problems Selçuk Bayraktar and his family faced for the sake of their ideals, I believe it would be beneficial to explain how the faith that led them to their path developed, as well as the motivational factors behind their success. Selçuk Bayraktar spent his childhood as an apprentice in an industrialist family producing complex automotive spare parts. He worked with his father Özdemir Bayraktar programming CNC machines in the workshop, which piqued his interest in robotics science. Notably, from an early age he had grown up in a productive industrialist culture, just like the garage culture in the USA. As a result, he had gained capabilities and skills in this field. His curiosity about aviation also started when his father Özdemir Bayraktar took him to a pilot course at the Turkish Air Association, where Vecihi Hürkuş also worked at certain times. Military demonstration flights on national holidays had particularly excited and encouraged Selçuk Bayraktar since childhood. What started with making model planes in his youth, shaped and led him to construct armed, uncrewed aerial vehicles (UAVs) with jet engines today. These material conditions effectively molded Selçuk Bayraktar's technical background. In addition to these, his family was influential in his spiritual-metaphysical growth that had shaped his national consciousness, just as Vecihi Hürkuş had promoted decades ago.

The greatest motivational factor for Selçuk Bayraktar, who grew up in a nationalist-conservative milieu, has been Turkish-Islamic civilization. Bayraktar considers the destruction of Takiyüddin Observatory based on an ill omen[37] during the Ottoman era as a significant breakdown in the relationship between Turkish-Islamic civilization and science. However, he believed religious motivation to still be effective in the West, and stated that an engineer, whom he had met in America and who had designed the Comanche and Apache helicopters at Boeing company, had worked fervently because Jesus was a carpenter.[38]

Bayraktar worked with foreign scientists from all over the world at Fort Benning Infantry Base in the early 2000s, and the question on his mind, and one supported by the authorities there, was whether such an environment could be provided in Türkiye. In fact, the idea of civil or military bureaucrats accrediting a Turkish university, let alone foreign researchers being invited to work at an army base outside of working hours, at that time were replete with bureaucratic obstacles and penalties that could end his career or even ruin his life. The bureaucratic structure did not even permit civilians to enter, let alone work, at the base. Moreover, when he asked a retired non-commissioned officer from the US Special Forces responsible for the administrative affairs and needs of the students there, the answer he received was: "If we don't give you this opportunity to work, even though your work is too meager to contribute to us at the moment, they will close this place as unproductive and useless". This allowed him to comprehend how the United States had utilized foreign brain power to national advantage and benefit within the military-industrial complex.[39]

Indeed, many Turkish, Thai, French, Italian, and Israeli students have stayed in the United States. Selçuk Bayraktar's frequent trips to Türkiye while studying for his doctorate at MIT also attracted the attention of his teachers at the school. He vividly remembers his Greek-American teacher half-jokingly approaching him, saying: "We are awarding you a scholarship, but not so that you may return to your homeland and build an aircraft". In the end, he did return to Türkiye, as the uncrewed aerial vehicle initiatives he was carrying out with his family in Türkiye had become inoperable without being there in person.[40]

In 2005, the Turkish Undersecretariat of Defense Industry (SSM) announced a national project, albeit an insignificant and small one, and decided to make a MINI UAV independent of foreign software. Although the authorities at the top of the decision-making mechanism had developed the vision of "We must and can produce our weapons indigenously", transformation in the halls of bureaucracy has never been quick and painless. When Selçuk Bayraktar went to meet with the Undersecretariat of Defense Industry within the framework of this project, the advice of the competent bureaucrat there was remarkable:

> You seem to be an intelligent and well-educated person, brother. Foreigners are technologically advanced in this area. Therefore, do not concern yourself with these matters. However, facilitate our contact with foreigners by acting as an interpreter for us and ensuring that we understand them. This is enough for us. You have a greater grasp of their language.[41]

From the perspective of discourse analysis, not only did this mid- or low-level bureaucrat lack confidence in the feasibility of domestic production in this field, but he also lacked awareness of the vision that had been adopted at the upper echelon. Maybe *prima facie* jealousy or material interests had led to a project supporting domestic production being suspended. However, Bayraktar produced the Mini UAV with his own resources and family without R&D support; in other words, just like Hürkuş had done in the workshop he had rented in Kadıköy. In the Mini UAV Development Project competition, BAYKAR's Mini UAV prototype was the only one to have a domestic robotic structure and software, unlike the other three foreign dependent participating companies; it also managed to fly, contrary to the prevailing belief in 2005. The foreign UAVs either failed to take off or crashed after takeoff. As Selçuk Bayraktar emphasized, the reason for the bureaucrats' astonishment at that moment, apart from financial reasons, was the "learned helplessness" that had probably originated from bitter past experiences. The question of "how come those from abroad could not fly yet the domestic one could" [42] comes to mind, and it derives from the "peripheral country" syndrome.[43]

Despite domestic production having been included as a term in the contract, the BAYKAR UAV's cheaper cost, and the others' inability to fly, members of the bureaucracy still could not decide in the face of this unexpected result. As a result, some military bureaucrats who were active in these processes delayed this project for about a year. However, the Mini UAV successfully completed its experimental flights in 2006 and managed to enter military inventory in 2007.[44] When comparing Bayraktar's Mini UAV with its peers, the US-made Raven and Israeli-made Skylark, the Mini UAV is seen to have twice the flight range and over three times the operational altitude.[45] Yet, it costs around 14% of the price of its foreign counterparts.

The success of the Mini UAV resulted in the bureaucracy primarily moving from behavioral patterns of ignorance and compromise, through concession, to the stage of obstruction. Known as the Malazgirt *Döner Kanat* [Rotary Wing], the Mini UAV's development began within the 6th Motorized Infantry Brigade of Şırnak with the approval of General Hasan Iğsız. The deficiencies identified in test flights were fixed a year later, and demonstration flights with two Malazgirt Mini UAVs were made for the bureaucrats of the Undersecretariat of Defense Industry in 2007. The fact that these prototypes, produced with the equities of BAYKAR Tech, passed the approval tests enabled the Turkish Land Forces Command to order four Malazgirt Mini UAVs in 2008. In parallel with this process, the project for transforming H300 helicopters, which had been removed from inventory, into uncrewed systems free-of-charge with the proposal from BAYKAR Tech and approval from the 2nd Chief of Defense General Ergin Saygun within the 5th Main Maintenance Center, was rejected for "not being cost-effective."

Despite the approval of the Presidency of General Staff with its higher bureaucratic authority, the structure that Vecihi Hürkuş had portrayed as an establishment gnawing on Turkish aviation from the inside managed to block the project. A year later, Malazgirt Mini UAVs flights that had already started operating on frontlines were ceased by the Technical Department of the Land Forces Command due to a "software error". Even though BAYKAR Tech officially guaranteed that a new one would be provided irrespective of the reason for the aircraft incident, Malazgirt UAV did not receive another flight permit.[46] In 2010, a domestic company that had been established as a result of the American arms embargo during the 1970s collaborated with the American company Guided Systems Technologies and bought an uncrewed helicopter system[47], and even recruited one of the software engineers working on BAYKAR Tech's R&D team.[48] However, the uncrewed helicopter project failed[49] and led to a public loss.

As seen in the example above, national culture and moral values, which Selçuk Bayraktar constantly emphasized during our interview with him, are more important than technical competence in military transformation. He mentioned that generals in the upper military bureaucracy such as Hasan Iğsız, Ergin Saygun, Saldıray Berk, Işık Koşaner, Necdet Özel, and Arif Çetin had given great support to his national defense industry projects, despite his father Özdemir Bayraktar being close to the *Milli Görüş*, a kind of Islamic political movement.[50] As seen in the example of President Mustafa Kemal Atatürk and General Chief of Staff Marshal Fevzi Çakmak giving open support to Vecihi Hürkuş for developing the national aviation industry, a patriotic stance, determination, and vision at the highest levels are not enough for success. The bureaucracy is a dominant power that causes moves to be interrupted due to its non-innovative structure, the jealousy from and financial interests of individuals in the system, or even simply their ignorance.

Despite the bureaucratic obstacles, BAYKAR Tech continued to work on technology development with civil-military cooperation in the field, instead of swerving from their values and true path. In 2007, they started to develop tactical UAVs using their own resources. The whole family, including their mother, Canan Bayraktar, participated in these activities around the clock in the factory and continued their lives by focusing on these works. In the scope of the Tactical UAV Development Project held by the Undersecretariat of Defense Industry in 2009, just as in the case of Vecihi Hürkuş, interest groups and their bureaucratic extensions employed the weapons of discrediting and sabotage as the last step. The BAYKAR-made MALE UAV's ability to automatically land and take off during a test flight carried out in the same year, something that other companies could not do at that time, almost led to Selçuk Bayraktar's brother Haluk Bayraktar being arrested. The extensions of interest groups in the bureaucracy claimed that the Tactical UAV had landed manually, igniting a quarrel. According to Selçuk Bayraktar, Israeli pilots landed the Herons manually. The reason behind this overreaction was revealed after three months. During the argument, the officer who had tried to arrest Haluk Bayraktar later resigned and accepted a lucrative position with a Turkish company associated with a foreign partner.[51]

Afterward, Selçuk Bayraktar told us how the bureaucracy had added a new article to the contract overnight, stating that the UAV is to be "launched from a catapult and land with a parachute" in order to integrate the other company into the system. An agreement was signed with the other company for six UAVs that could not fly. Having high costs due to its use of a for-eign-made flight control systems, the company then reduced its previous overtly high-cost offer to compete with BAYKAR's cost, which was two-thirds cheaper. As Özdemir Bayraktar said, "despite this company's aircraft being unable to fly, they managed to fly very well in the corridors of bureau-cracy".[52] As a result, the timeline for the production phase of this company's UAVs was extended, which opened the way for BAYKAR's Bayraktar TB2, which was produced with equity capital, passed all tests, and started mass production in 2014. Despite the first delivery having been made to the Turkish Armed Forces in the same year, the MALE UAVs were kept in hangars for ten months; this weapon that could create an asymmetry in the fight against terrorism was kept grounded, despite the high losses resulting from the ter-rorist organization PKK/KCK's raids on military bases. In other words, bureaucratic obstacles experienced at the production and delivery stages con-tinued at the later stages.

At this point, one should not forget that a significant amount of time is required for new technological systems to cause a change in large organiza-tions such as armies. Although free training was provided for nine months with the encouragement of BAYKAR Tech, the process of developing a national doctrine was more complex and took longer, just as in other armies, which led to the ineffective use of UAVs.[53] The scarcity of weapons bought from abroad over the past 200 years has additionally resulted in their overval-uation, which has in turn increased the bureaucratic barriers to weapon usage. In the end, no one was willing to crash one of the increasingly scarce UAVs. As a result, it took time to overcome the overwhelming fear of aircraft loss.[54]

In 2015, a proposal came up again to lease the aircraft that had been pur-chased the previous year. As Bayraktar stated, the bureaucracy offered an "incentive" in order to involve the other company in the game again. The bureaucracy was offering money to the manufacturer through re-leasing for aircraft that had already been bought and paid for. Selçuk Bayraktar clearly remembers one of the officials saying, "If you don't lease it, it does not [can-not] enter the inventory", despite BAYKAR Tech saying they considered it unethical and contrary to their beliefs.[55]

Overcoming bureaucratic obstacles in an ecosystem dominated by favorit-ism involving global-scale companies and Turkish collaborators with foreign partners and interest groups was made possible through what Selçuk Bayraktar called the "third way". As he had said before, setting out in line with an idealist belief and a "sacred" ideal and not giving up in the face of any kind of threat, blackmail, or sabotage ensures success. From a technical point of view, being in the field from the beginning had enabled needs to be determined on-site and deficiencies to be detected and corrected immediately, just as Vecihi Hürkuş had done. In Selçuk Bayraktar's own words, no official or legal entity had told him to "make a drone and arm it".[56]

Soldiers usually do not allow civilians to get too involved in their affairs; many simply do not like it. However, the General Manager of BAYKAR Tech at the time, Özdemir Bayraktar at first, followed by Haluk and Selçuk Bayraktar and their engineers, worked shoulder to shoulder voluntarily with the soldiers at the front. They stayed together in the barracks, sharing the same meals and, more importantly, observing the needs of the soldiers first hand, creating trust between soldiers and civilians and ensuring the development of a technology capable of being used most accurately on the battlefield. While this "amateur" civil-military collaboration may be taken for granted in industrialized Western nations, it is a novel thing in Türkiye. This type of civil-military collaboration created asymmetry for BAYKAR Tech, as companies with great financial power on a global scale or that enter the defense industry with the expectation of pure financial gain have different motivations than theirs. Selçuk Bayraktar stated that he spent about four years in high-risk areas at various times and had worked at almost all outposts and bases.

In his own words, "Working with the soldiers in the field provided everyone with a spiritual transformation, from the engineers to the bureaucrats who took part in that struggle, and proving themselves every day has helped break down bureaucratic obstacles". For example, ROKETSAN also worked with BAYKAR Tech for about two years in the field and contributed to equipping UAVs with domestic ammunition. Although military bureaucracy did not allow firing tests at first and the trials had been prohibited, permission was obtained as a result of political initiatives.[57] By the end of this process, mini smart ammunition (MAM) has been exported to many countries, including buyers in Europe, and is actively being used in several conflicts.[58]

In Lieu of a Conclusion: The Turkish UAVs in Action

The development of UAVs is not a process that can be explained just by having sufficient technical and scientific knowledge. The information that Bayraktar's family personally relayed has revealed the combination of political, military, economic, and cultural factors to be imperative. First of all, one may safely assume that the family embraced the work as a "sacred cause" and persevered with remarkable faith. Secondly, the family chose the path of developing on-site, leaving their comfort zones and sharing the lived experience of those in need (i.e., the soldiers on the frontline). This method may be considered an ordinary phenomenon in developed and industrialized countries. However, it is an essential criterion for countries such as Türkiye, which until recently had volatile civil-military relations hindering effective civil-military cooperation. The defense industry has additionally been heavily dependent on foreign power, leaving no room for partnership between civilian and military experts. A third factor is that support from political and military elites was provided to break the monopoly of a bureaucracy resistant to change as well as the monopoly of countries in the defense industry. Moreover, this support was granted irrespective of any political or ideological discrimination. Therefore, it could be suggested that Türkiye has learned lessons from its past, relatively painful, experiences.

In fact, the painful, long UAV production process has finally begun to bear fruit. Even though UAVs alone are not enough to win a war in today's conflicts[59], they do undeniably create considerable asymmetry. For example, UAVs in Idlib were able to destroy 20 tanks that terrorists had captured from a Syrian tank division and sent to the Turkish border. Small 600 kg UAVs are only a third of the cost of a tank, and they destroyed the tanks lined up like prayer beads with their small but smart bombs, as quickly as a shoal of piranhas attacking a shark. Air defense systems that have made their impact felt from the last quarter of the Cold War to the present day have also been ineffective against this new generation of small but intelligent weapons both in the Libyan Civil War and the Nagorno-Karabakh War, in which Azerbaijan liberated territories Armenians had occupied for 25 years. Turkish UAVs have destroyed Russian-made SA-15 Tor, Pantsir-S1, and even S300 air defense missile systems.[60] UAVs will likely be able to also defeat crewed aircraft in the near future, if not today.

The UAVs have been used against the terrorist organization PKK/KCK before conventional conflicts, have provided an excellent advantage for Türkiye in fighting terrorism, and have been able to silence disinformation campaigns both at home and abroad. For instance, after the UAVs destroyed a target in Beytüşşebbap at an altitude of 10,600 feet on September 29, 2016, the UAVs were claimed to have killed shepherds walking with their herd or innocent villagers having a picnic. This claim was even repeated in parliament and on social media, putting Türkiye under pressure from the public. However, the black propaganda was reversed after the images the UAVs had taken of the operation were presented to the national and international press. These so-called "citizens" were found to have excavated trenches at 10,600 feet and to possess heavy machine guns (DShK) and radio antennas for communicating with distant points such as Mount Kandil, where the terrorist organization PKK/KCK is based.

UAVs have arisen from a pressing need for counter-terrorism operations in Türkiye and have also been used by different countries in peace-support operations and regional conflicts. Even though the American and Israeli armies, the Russians, and the Chinese have used them in single-precision strikes and covert operations for many years, this technology was actively integrated into conventional war by the Turkish Army. These new drone operations have also triggered cultural change that has helped break down the vicious cycle of learned helplessness.

Notes

1 For detailed information: Max Weber, *Grundriss der Sozialökonomik, III. Abteilung, Wirtschaft und Gesellschaft*, J.C.B. Mohr (Paul Siebeck), Tübingen, 1922, pp. 696–697.
2 Weber, *Grundriss der Sozialökonomik*, pp. 698–700 and pp. 703–707.
3 Graham T. Allison, "Conceptual Models and the Cuban Missile Crisis", *The American Political Science Review*, 63:3, 1969, pp. 707–708.

4 Allison, "Conceptual Models and the Cuban Missile Crisis", p. 709.
5 Michel Foucault, *Büyük Kapatılma- Seçme Yazılar 3*, (Transl. Işık Ergüden, Ferda Keskin), Ayrıntı Yayınları, İstanbul, 2007, passim.
6 İsmail Yavuz, *Mustafa Kemal'in Uçakları, Türkiye'nin Uçak İmalat Tarihi (1923–2012)*, Türkiye İş Bankası Yayınları, İstanbul, 2015, pp. 11–12.
7 Vecihi Hürkuş, *Havalarda 1915–1925*, Ahmet Sait Basımevi, İstanbul, 1942, p. 45.
8 Hürkuş, *Havalarda 1915–1925*, p. 46.
9 Emait: It is a material used in the replacement and various maintenance of cloths of old aircraft, wings of which were made of cloth.
10 Hürkuş, *Havalarda 1915–1925*, pp. 48, 134–138.
11 The engine of this type of aircraft is 80 hp, its speed is 180 km, and the full load take-off weight is 890 kg. Yavuz, *Mustafa Kemal'in Uçakları, Türkiye'nin Uçak İmalat Tarihi (1923–2012)*, pp. 14–18.
12 Hürkuş, *Bir Tayyarecinin Anıları*, (eds.) Gönül Hürkuş Şarman, Sevim Hürkuş Maxon, Yapı Kredi Yayınları, İstanbul, 2021, p. 224.
13 Hürkuş, *Bir Tayyarecinin Anıları*, p. 228.
14 For the support of the executive elite, see Hürkuş, *Bir Tayyarecinin Anıları*, p. 232–233.
15 Hürkuş, *Bir Tayyarecinin Anıları*, p. 248.
16 Osman Yalçın, *Türk Hava Harp Sanayii Tarihi*, Türkiye İş Bankası Yayınları, İstanbul, 2013, p. 193.
17 Nuri Demirağ (1886–1957): Demirağ, one of the businessmen of the early Republican period, broke ground in Turkish industries such as domestic cigarette paper production and domestic parachute production, and completed more than 1000 km of railways in one year, also undertaking the construction of a number of state-owned factories. In 1931, he developed a bridge project similar to the Golden Gate Bridge in San Francisco over the İstanbul Strait, but although the project was accepted by the president of that period, Mustafa Kemal Ataturk, it did not receive approval from the government. Demirağ, whose dam project on Keban was also rejected, became interested in aviation, and, like Vecihi Hürkuş, opened an aircraft factory and aviation schools, one in Sivas as a secondary school and the other in İstanbul Yeşilköy as a high school named Gökokul in 1936. In the same year, a single-engine trainer aircraft named "Nu.D-36" was produced in the building where the Shangri-La Hotel is located in Beşiktaş today, and a twin-engine six-seat airliner named "Nu.D-38" were produced in the factory in Yeşilköy in 1938. In 1941, Demirağ bought a large plot of land in Yeşilköy to conduct flight tests of aircraft, and soon got a large flight site, hangars, and aircraft repair shop built, enabling it to be used as the largest international airport in İstanbul until 2019. Demirağ's aircraft factory, which eventually sold 65 gliders to the Turkish Air Association, was initially defeated by the prevailing bureaucracy. After a Nu.D-36 aircraft was involved in a fatal accident during landing, the Turkish Air Association canceled orders and a law to prevent sales abroad was passed. After the court's decision in favor of the Turkish Air Association, Demirağ, by taking advantage of the switch to a multi-party system, founded the National Development Party, which was the first opposition party of the republican period. Demirağ, who entered politics "in search of justice", attracted more attention from the bureaucracy, and all his aviation-related assets and the school that raised 290 pilots were expropriated. Demirağ was elected as a deputy in the Democratic Party in 1954 after a failed opposition party attempt, and passed away in 1957. Fatih M. Dervişoğlu, *Nuri Demirağ, Türkiye'nin Havacılık Efsanesi*, Ötüken Yayınları, İstanbul, 2007.
18 On October 10, 1940, Vecihi Hürkuş could receive his diploma in engineering and mechanical engineering by the decision of the Council of State No. 40/211 Hürkuş, *Bir Tayyarecinin Anıları*, p. 382.
19 Dervişoğlu, *Nuri Demirağ, Türkiye'nin Havacılık Efsanesi*, p. 109.

20 "NSC 162/2, Basic National Security Policy, October 30, 1953, Washington, https://fas.org/irp/offdocs/nsc-hst/nsc-162-2.pdf, p. 21, accessed on 01.05.2021.

21 Hürkuş, *Bir Tayyarecinin Anıları*, pp. 387–397.

22 Hürkuş, *Bir Tayyarecinin Anıları*, pp. 398–399, 412–413.

23 The script for the movie named *Gülen Gözler*, directed by Ertem Eğilmez, written by Sadık Şendil and produced by Nahit Ataman. It was set in the Yeşilköy district of İstanbul, where the airport is located. "Gülen Gözler" https://www.imdb.com/title/tt0289967/, accessed on 04.07.2021.

24 Kaan Kutlu Ataç and Mehmet Mert Çam, "Soğuk Savaş: Amerikan Büyük Stratejisi, Johnson Doktrini ve Kıbrıs", *Journal of Security Strategies*, 16:35, 2020, pp. 627–628.

25 Ataç and Çam, "Soğuk Savaş: Amerikan Büyük Stratejisi, Johnson Doktrini ve Kıbrıs", pp. 623–624 and for the original of the letter, see "Johnson Letter", https://www.scribd.com/document/38718043/Johnson-Letter, accessed on 19.04.2021.

26 Henry Kissinger saw the area of influence of the Soviets in the Republic of Cyprus. In both Greece and the Cyprus republic, coups that disrupted the "order" were also effective in this regard. For Kissinger's opinion, see *Foreign Relations of the United States (FRUS), 1969–1976, Volume XXX, Greece; Cyprus; Turkey, 1973–1976*, (eds.) Laurie Van Hook, Edward C. Keefer, US Government Printing Office, Washington, 2007, pp. 423–424.

27 "Turkey Visited by Jinsa Delegation", https://jinsa.org/jreport/turkey-visited-by-jinsa-delegation/, accessed on 19.04.2021.

28 The modernization of helicopters specifically turned into a problematic experience. Two of the 19 helicopters bought in 1992 crashed in Hakkari, the manufacturer Rosoboronexport was replaced by Kazan Helicopters, and lawsuits were filed against the new manufacturer on charges of tax evasion, and four helicopters disappeared in Russia. Then it was decided that the avionics modernization of helicopters should be carried out by ASELSAN. "4 helikopterimiz Rusya'da kayboldu" https://www.hurriyet.com.tr/gundem/4-helikopterimiz-rusyada-kayboldu-7338260, accessed on 19.04.2021. Then the Russians stepped in again. "Rusya, Mi-17 helikopterlerini Türkiye'de onarabilir" https://tr.sputniknews.com/rusya/201505081015379562/, accessed on 19.04.2021. Finally, it was agreed with Ukraine. "Jandarma'nın Mi-17'lerini Ukrayna modernize edecek" https://www.defenceturk.net/jandarma-mi-17-ukrayna-modernize, accessed on 19.04.2021.

29 In 2013, the United States bought 30 Mil Mi-17 helicopters for use in high and hot areas such as Afghanistan. "U.S. Army to buy 30 Russian Mi-17 helicopters for use in high, hot areas of Afghanistan" https://www.militaryaerospace.com/communications/article/16715565/us-army-to-buy-30-russian-mi17-helicopters-for-use- in-high-hot-areas-of-afghanistan, accessed on 19.04.2021.

30 Turkey Visited by Jinsa Delegation", https://jinsa.org/jreport/turkey-visited-by-jinsa-delegation/, accessed on 19.04.2021.

31 Efraim Inbar, *Türk-İsrail Stratejik Ortaklığı*, (Transl. Suna Ercan, Özgül Erdemli), Eurasian Center for Strategic Studies (ASAM), Ankara, 2002, passim.

32 "İsrail'den alınan Heronlar çürük çıktı" https://www.ntv.com.tr/turkiye/israilden-alinan-heronlar-curuk-cikti, XUm9bUNj80yIfKAzftDBEA, accessed on 19.04.2021.

33 "Kayıp Heron'u İsrail buldu", https://www.yenisafak.com/gundem/kayip-heronu-israil- buldu-273997, accessed on 19.04.2021.

34 "Casus uçak mı açık hedef mi", https://www.yenisafak.com/gundem/casus-ucak-mi-acik-hedef-mi-154354, accessed on 20.04.2021.

35 *İnsansız Hava Aracı*, Ankara, Training and Doctrine Command (EDOK), 2000, pp. 4–18.

36 Interview with Özdemir, Selçuk and Haluk Bayraktar by the author March 2021, BAYKAR Tech, İstanbul, 6 March 2021.

37 Takiyuddin also researched in mathematics, and prepared trigonometric rulers, as one of the architects of the early European Enlightenment, and made definitions

and practical proofs of them. Takiyuddin, designing mechanical watches that could show minutes and seconds, also calculated reflection angles in the field of optics and formulated them in his works. The comet seen in İstanbul for one month and the epidemic and earthquake that occurred immediately after were attributed to the observatory and Takiyuddin's work. Murat III, who failed to achieve the results expected from the Iranian Expedition, destroyed the observatory with a cannon shot from the sea in 1577 with the fatwa of Shayk al-İslam stating that "it is inappropriate to study the secrets of angels", although he did not believe in superstitions. Mustafa Kaçar et al. *XVI. Yüzyıl Osmanlı Astronomu Takiyüddin'in Gözlem Araçları – Âlât-ı Rasadiyye li Zîc-i Şehinşâhiyye*, İstanbul, Türkiye İş Bankası Kültür Publishing, 2011. For the works of Takiyuddin bin Maruf-i, see Remzi Demir, *Takiyüddin'de Matematik ve Astronomi Ceridetü'd-Dürer ve Haridetü'l-Fiker Üzerine Bir İnceleme*, Atatürk Culture Center, Ankara, 2000, passim.

38   Interview with Selçuk Bayraktar, March 2021, BAYKAR Tech, İstanbul.

39   Military Industrial Complex: A theory suggested by C. Wright Mills in 1956 and mentioned by the President of the United States, Dwight D. Eisenhower on 17 January 17, 1961 in his farewell speech. The theory explains the directive effect of the sophisticated relationship between the armed forces in America and the defense industry that provides it with supplies, which has interests on both sides of public policy. "Military Industrial Complex" https://www.britannica.com/topic/military-industrial- complex, accessed on 20.04.2021. Eisenhower implies that the military-industrial complex is vital to survive, but warns that this structure, which directly affects the social structure, should not be allowed to jeopardize freedoms and democratic processes, and violate the law. "Military-Industrial Complex Speech", Dwight D. Eisenhower, 1961, https://avalon.law.yale.edu/20th_century/eisenhower001.asp, accessed on 20.04.2021.

40   Interview with Selçuk Bayraktar, 6 March 2021, BAYKAR Tech, İstanbul.

41   Interview with Selçuk Bayraktar, 6 March 2021, BAYKAR Tech, İstanbul.

42   Interview with Selçuk Bayraktar, March 2021, BAYKAR Tech, İstanbul.

43   Peripheral Country: In the "Modern World System Theory" in which Immanuel Wallerstein treated the financial dependence among countries he classified in two main groups as "central: and "peripheral" Central countries produce expensive and high technology products with cheap raw materials, agriproducts and cheap labor provided by peripheral countries, and then sell them back to peripheral countries. Immanuel Wallerstein, *The Modern World-System: Capitalist Agriculture and the Origins of the European World-Economy in the Sixteenth Century*, Academic Press, New York, 1976, pp. 343–353. Peripheral Country Syndrome, as is understood by me, that a peripheral country's citizens may incline to believe that 'the product purchased from abroad is always better for the society of a country where there is no capitalist structure connected with production relations and that it can never produce it itself'.

44   "Bayraktar Mini UAV", https://baykardefence.com/uav-16.html, accessed on 30.04.2021.

45   "Bayraktar Mini İHA, Raven ve Skylark Mini İHA Karşılaştırması" https://twitter.com/savunma__sanayi/status/926830948250927104, accessed on 30.04.2021.

46   Interview with Selçuk Bayraktar, March 2021, BAYKAR Tech, İstanbul.

47   There is a Turkish company on the "customers" tab of the company's website called Guided Systems Technologies. "Customers", http://guidedsys.com/customers/, accessed on 01.05.2021. However, cooperation in the status of "technology transfer" is not included in the Turkish company's annual report, and the name of the American company is not mentioned. "2010 Yılı Faaliyet Raporu" https://www.aselsan.com.tr/2010_Faaliyet_Raporu_6686.pdf, accessed on 01.05.2021.

48   "Tüfek yapmakta geç kaldık, bari milli İHA vurulmasın!" https://www.haberturk.com/yazarlar/guntay-simsek-1019/1038712-tufek-yapmakta-gec-kaldik-bari-milli-iha-vurulmasin, accessed on 01.05.2021.

49 "Nizip'te İnsansız Hava Uçağı Paniği" https://www.hurriyet.com.tr/gundem/ nizipte-insansiz-hava-ucagi-panigi-26387458, accessed on 01.05.2021.
50 Interview with Selçuk Bayraktar, 6 March 2021, BAYKAR Tech, İstanbul.
51 Interview with Selçuk Bayraktar, 6 March 2021, BAYKAR Tech, İstanbul.
52 "İnsansız hava aracı geliştirme projesi imzalandı" https://www.milliyet.com.tr/ ekonomi/insansiz-hava-araci-gelistirme- projesi-imzalandi-1334827, accessed on 01.05.2021. Interview with Özdemir Bayraktar, March 2021, BAYKAR Tech, İstanbul.
53 Developing a national doctrine in line with a domestically produced weapon system, refers to a longer, more challenging, and sophisticated process when considered in terms of actors involved, such as military and civilian experts, academics, and political figures.
54 The process taking place in Herons confirms this statement of mine. After the Gaza flotilla raid in 2010, the small number of Herons' flying at low altitudes so that they did not crash, which had repair and modernization problems, was banned.
55 Interview with Selçuk Bayraktar, 6 March 2021, BAYKAR Tech, İstanbul.
56 Interview with Selçuk Bayraktar, 6 March 2021, BAYKAR Tech, İstanbul.
57 Interview with Selçuk Bayraktar, 6 March 2021, BAYKAR Tech, İstanbul.
58 "5 binden fazla MAM-L ve MAM-C teslim edildi" https://www.defenceturk. net/5-binden- fazla-mam-l-ve-mam-c-teslim-edildi, accessed on 01.06.2021.
59 Conn Hallinan, "Day of The Drone: We Need an International Convention on Drones" https://fpif.org/day-of-the-drone-we-need-an-international-convention- on-drones/, accessed on 02.05.2021. John V. Parachini, Peter A. Wilson "Drone- Era Warfare Shows the Operational Limits of Air Defense Systems", https:// www.rand.org/blog/2020/07/drone-era-warfare-shows-the-operational-limits-of- air.html, accessed on 02.05.2021.
60 "Turkish Bayraktar TB2 drone destroyed the S-300 SAM of Armenia" https:// www.globaldefensecorp.com/2020/10/08/breaking-news-turkish-baytraktar-tb-2- destroyed-the-s-300-sam-of-armenia/, accessed on 02.05.2021. "TSK tarafından imha edilen Rus yapımı Rejim Pantsir-S1 hava savunma sistemi" https://www. defenceturk.net/tsk-tarafindan-imha-, [accessed on 02 May 2021]. "Libya'daki Bayraktar TB2 İHA'lar ve Rus "Pantsirler" https://haber.aero/aero-gundem/ libyadaki-bayraktar-tb2-ihalar-ve-rus-pantsirler/, accessed on 02.052021.

References

"2010 Yılı Faaliyet Raporu" https://www.aselsan.com.tr/2010_Faaliyet_Raporu_6686. pdf, accessed on 01.05.2021.

"4 helikopterimiz Rusya'da kayboldu" https://www.hurriyet.com.tr/gundem/4- helikopterimiz-rusyada-kayboldu-7338260, accessed on 19.04.2021.

"5 binden fazla MAM-L ve MAM-C teslim edildi" https://www.defenceturk.net/5- binden- fazla-mam-l-ve-mam-c-teslim-edildi, accessed on 01.05.2021.

Allison, T. Graham, "Conceptual Models and the Cuban Missile Crisis", *The American Political Science Review*, 63:3, 1969, pp. 689–718.

Ataç, Kaan Kutlu and Çam, Mehmet Mert, "Soğuk Savaş: Amerikan Büyük Stratejisi, Johnson Doktrini ve Kıbrıs", *Journal of Security Strategies*, 16:35, 2020, pp. 595–634.

Bayraktar, Haluk, Interview by Mehmet M. Çam, 6 March 2021a, BAYKAR Tech, İstanbul.

"Bayraktar Mini İHA, Raven ve Skylark Mini İHA Karşılaştırması" https://twitter. com/savunma__sanayi/status/926830948250927104, accessed on 30.04.2021.

"Bayraktar Mini UAV", https://baykardefence.com/uav-16.html, accessed on 30.04.2021.

Bayraktar, Özdemir, Interview by Mehmet M. Çam, 6 March 2021b, BAYKAR Tech, İstanbul.

Bayraktar, Selçuk, Interview by Mehmet M. Çam, 6 March 2021c, BAYKAR Tech, İstanbul.

"Casus uçak mı açık hedef mi", https://www.yenisafak.com/gundem/casus-ucak-mi-acik-hedef-mi-154354, accessed on 20.04.2021.

Conn, Hallinan, "Day of The Drone: We Need an International Convention on Drones" https://fpif.org/day-of-the-drone-we-need-an-international-convention-on-drones/, accessed on 02.05.2021.

"Customers", http://guidedsys.com/customers/, accessed on 01.05.2021.

Demir, Remzi, *Takiyüddin'de Matematik ve Astronomi Ceridetü'd-Dürer ve Haridetü'l-Fiker Üzerine Bir İnceleme*, Atatürk Culture Center, Ankara, 2000.

Dervişoğlu, M. Fatih, *Nuri Demirağ, Türkiye'nin Havacılık Efsanesi*, Ötüken Yayınları, İstanbul, 2007.

Foreign Relations of the United States (FRUS), 1969–1976, Volume XXX, Greece; Cyprus; Turkey, 1973–1976, (eds.) Laurie Van Hook, Edward C. Keefer, United States Government Printing Office, Washington, 2007.

Foucault, Michel, *Büyük Kapatılma- Seçme Yazılar 3*, (Transl. Işık Ergüden, Ferda Keskin), Ayrıntı Yayınları, İstanbul, 2007.

"Gülen Gözler" https://www.imdb.com/title/tt0289967/, accessed on 04.04.2021.

Hürkuş, Vecihi, *Bir Tayyarecinin Anıları*, (eds.) Gönül Hürkuş Şarman, Sevim Hürkuş Maxon, Yapı Kredi Yayınları, İstanbul, 2021.

Hürkuş, Vecihi, *Havalarda 1915–1925*, Ahmet Sait Basımevi, İstanbul, 1942.

Inbar, Efraim, *Türk-İsrail Stratejik Ortaklığı*, (Transl. Suna Ercan, Özgül Erdemli), Eurasian Center for Strategic Studies (ASAM) Yayınları, Ankara, 2002.

İnsansız Hava Aracı, Training and Doctrine Command (EDOK), Ankara, 2000.

"İnsansız hava aracı geliştirme projesi imzalandı" https://www.milliyet.com.tr/ekonomi/insansiz-hava-araci-gelistirme-projesi-imzalandi-1334827, accessed on 01.05.2021.

"İsrail'den alınan Heronlar çürük çıktı" https://www.ntv.com.tr/turkiye/israilden-alinan-heronlar-curuk-cikti,XUm9bUNj80yIfKAzftDBEA, accessed on 19.04.2021.

"Jandarma'nın Mi-17'lerini Ukrayna modernize edecek" https://www.defenceturk.net/jandarma-mi-17-ukrayna-modernize, accessed on 19.04.2021.

"Johnson Letter", https://www.scribd.com/document/38718043/Johnson-Letter, accessed on 19.04.2021.

Kaçar, Mustafa et al. *XVI. Yüzyıl Osmanlı Astronomu Takiyüddin'in Gözlem Araçları – Âlâtı Rasadiyye li Zîc-i Şehinşâhiyye*, Türkiye İş Bankası Kültür Yayınları, İstanbul, 2011.

"Kayıp Heron'u İsrail buldu", https://www.yenisafak.com/gundem/kayip-heronu-israil-buldu-273997, accessed on 19.04.2021.

"Libya'daki Bayraktar TB2 İHA'lar ve Rus "Pantsirler"" https://haber.aero/aero-gundem/libyadaki-bayraktar-tb2-ihalar-ve-rus-pantsirler/, accessed on 02.05.2021.

"*Military-Industrial Complex Speech*", Dwight D. Eisenhower, 1961, https://avalon.law.yale.edu/20th_century/eisenhower001.asp, accessed on 20.04.2021.

"Nizip'te İnsansız Hava Uçağı Paniği" https://www.hurriyet.com.tr/gundem/nizipte-insansiz-hava-ucagi-panigi-26387458, accessed on 01.05.2021.

"NSC 162/2, Basic National Security Policy", October 30, Washington, 1953, https://fas.org/irp/offdocs/nsc-hst/nsc-162-2.pdf, accessed on 01.05.2021.

Parachini, John V., Peter A. Wilson, "Drone-Era Warfare Shows the Operational Limits of Air Defense Systems", https://www.rand.org/blog/2020/07/drone-era-warfare-shows-the-operational-limits-of-air.html, accessed on 02.05.2021.

"Rusya, Mi-17 helikopterlerini Türkiye'de onarabilir" https://tr.sputniknews.com/rusya/201505081015379562/, accessed on 19.04.2021.

"TSK tarafından imha edilen Rus yapımı Rejim Pantsir-S1 hava savunma sistemi" https://www.defenceturk.net/tsk-tarafindan-imha- accessed on 02.04.2021.

"Tüfek yapmakta geç kaldık, bari milli İHA vurulmasın!" https://www.haberturk.com/yazarlar/guntay-simsek-1019/1038712-tufek-yapmakta-gec-kaldik-bari-milli-iha-vurulmasin, accessed on 01.05.2021.

"Turkey Visited by Jinsa Delegation", https://jinsa.org/jreport/turkey-visited-by-jinsa-delegation/, accessed on 19.04.2021.

"Turkish Bayraktar TB2 drone destroyed the S-300 SAM of Armenia" https://www.globaldefensecorp.com/2020/10/08/breaking-news-turkish-baytraktar-tb-2-destroyed-the-s-300-sam-of-armenia/, accessed on 02.05.2021.

"U.S. Army to buy 30 Russian Mi-17 helicopters for use in high, hot areas of Afghanistan" https://www.militaryaerospace.com/communications/article/16715565/us-army-to-buy-30-russian-mi17-helicopters-for-use-in-high-hot-areas-of-afghanistan, accessed on 19.04.2021.

Wallerstein, Immanuel, *The Modern World-System: Capitalist Agriculture and the Origins of the European World-Economy in the Sixteenth Century*, Academic Press, New York, 1976.

Weber, Max, *Grundriss der Sozialökonomik, III. Abteilung, Wirtschaft und Gesellschaft*, J.C.B. Mohr (Paul Siebeck), Tübingen, 1922.

Yalçın, Osman, *Türk Hava Harp Sanayii Tarihi*, Türkiye İş Bankası Kültür Yayınları, İstanbul, 2013.

Yavuz, İsmail, *Mustafa Kemal'in Uçakları, Türkiye'nin Uçak İmalat Tarihi (1923–2012)*, Türkiye İş Bankası Yayınları, İstanbul, 2015.

4 The Flagship of the Turkish Defense Industry

The Route to MİLGEM

Kemal Eker

Introduction

Sea, a massive component of geography, has molded lifestyles, economies, welfare, security perceptions, and international relations for centuries. Therefore, an adequate maritime power has been required for ruling the seas to the extent that the sea influences a state's interests. Mahan underlined the six determinants of naval power as: geographic location, physical structure, extent of territory, population size, character of the people, and character of government. Since Turks stepped into Anatolia, these conditions have propelled them to possess a maritime power and improve it, albeit not always successfully.

Turkish naval activities have been a point of interest, especially in recent years, primarily because of the tensions in the Mediterranean. However, exploring the development of Turkish naval engineering is crucial in order to better understand the current era. Therefore, this chapter will first summarize Turkish naval power and engineering history and then scrutinize the advent of the MİLGEM (National Warship) Project. Naval engineering triggers not just ship hull construction but also a significant number of sub-systems such as those involving engines and propulsion, command and control, weaponry, communications, and detection (i.e., radar and sonar). Consequently, building a ship such as MİLGEM using 65% domestic construction mobilizes the defense industry enterprises related to these sub-systems and contributes to their development.

The chapter's aim is to explain how the MİLGEM Project has made a breakthrough in the Turkish defense industry and become its flagship. The experiences gained throughout the project have also considerably affected both naval warfare ships and other fields in the defense industry. As an example, MİLGEM has resulted in a huge impact on İstanbul Class (I Class) frigates. In this regard, it is a milestone in the history of the Turkish defense industry and will play a substantial role how similar projects will be developed in the future.

A Brief History of Turkish Maritime Power and Industry

Some records show the Turks to have been engaged in shipping and fishing activities in the Caspian Sea, Lake Baikal, and rivers in Central Asia in

DOI: 10.4324/9781003327127-5

ancient times. However, saying that Turkish states were able to create a maritime power would be difficult, despite having ruled the Azov Sea and Black Sea coasts for a long time. The generally accepted thesis regarding the Turks' maritimization is that their strength in this field started with the conquest of Anatolia.[1] After prolonged struggles with the Byzantines, the Turks settled the people who were engaged in fishing and maritime activities in the closed seas of Turkestan along the Anatolian coasts they had conquered. Turkish maritimization emerged once the Oğuz sailors had set foot on Anatolian shores.[2]

The Anatolian Seljuks conquered İznik (Nikea) in 1075 and tried to capture Gemlik (Kios) in order to gain a foothold to the Sea of Marmara. However, the Seljuks' encounter with the strong Byzantine navy deflected their direction into the Aegean Sea.[3] The first Turkish Navy, composed of 17 galleys and 33 different sailboats, was organized by Çaka Bey who had employed shipbuilders from İzmir. This navy was launched on the Aegean Sea in 1081, taking advantage of the war between the Byzantines and Pechenegs in the Balkans.[4] This date is also accepted as both the foundation year of the Turkish Naval Forces and the start of Turkish naval war history.[5]

Even though the Crusades had obstructed the Anatolian Seljuks from gaining power in the seas in the following years, the Seljuks did manage to build ships in the Sinop and Alanya Shipyards and create a fleet during the reign of Sultan Aladdin Kayqubad I. Thus, the Ottoman Empire benefited from the experience and knowledge of these *beyliks* [territorial jurisdiction of a *bey*, or Ottoman governor] on maritime power. As a result, the Ottoman beyliks were able to get a foothold on the shore of Marmara, something the Anatolian Seljuks had failed to do, and conquer Karamürsel in 1323.[6]

The naval hierarchy was created following the establishment of the first Ottoman shipyard in Karamürsel in 1327, and the naval commandant was called *Derya Beyi* [Lord of the Sea]. Karamürsel Bey became the first Derya Beyi in the Ottoman Navy.[7] The first modern organization in the Ottoman Navy started under the reign of Bayezid I (Yıldırım Bayezid). Upon construction of the Gallipoli Naval Base in 1401, the title of *Kaptan-ı Derya* [Admiral in Chief] began being used for the naval commandant. Saruca Pasha became the first Kaptan-ı Derya of the Ottoman Navy.[8]

The progress of Turkish maritime power gained momentum during the reign of Mehmed the Conqueror. Following efforts to dominate the Aegean and Black Seas, the Ottoman navy set sail for the Mediterranean and gained power through the İstanbul Shipyard, founded in 1455 as the largest shipyard of the period.[9] As a result, Turkish admirals made substantial contributions to world maritime activities in the following years, the world maps drawn by Piri Reis and his naval studies (e.g., *Kitab-ı Bahriye*/The Book of the Navy) being some examples of these.[10]

After the conquest of Egypt by Selim I, the Ottoman Navy appeared in the Indian Ocean and the Red Sea. Suleiman the Magnificent added new ships to the Ottoman Navy. The victory of the Battle of Preveza under Kaptan-ı

Derya Barbaros Hayreddin Pasha (also known as Hayreddin Barbarossa) paved the way for the Ottomans to dominate the Mediterranean. In this period, the Turkish Navy was present in various seas thanks to great admirals such as Salih Reis, Piyale Pasha, Selman Reis, and Kılıç Ali Pasha. However, Ottoman statesmen have prioritized land power over sea power for reaching political goals since the 17th century. This consequently led to the decline of Turkish maritime activities, which then fell behind the West's technological advances. The transformation of ships into galleons similar to their peers in the West occurred late for the Ottomans.[11]

According to Fernand Braudel, the Ottoman Empire did not engage in any serious activity in the Mediterranean after the Battle of Lepanto in 1571 until the declaration of war against Venice in 1645 due to the Crete Island expeditions. Their rivals in this area had also given up the struggle. This period of peace appeared to be the beginning of the decline of Ottoman maritime activities due to naval inactivity. The navy no longer renewed or repaired those things that stopped working.[12]

In addition to Braudel's evaluation, the Lepanto defeat had also invalidated the idea of the invincibility of the Turks at sea. Apart from this psychological result, the fact that many knowledgeable, trained, and experienced sailors had lost their lives was also very significant. New ships were quickly built to replace those that had been lost, but taking up the slack from the sailors who had been lost was not possible.[13] The deaths of Kılıç Ali Pasha (1587) and Uluç Hasan Pasha (1590) in particular marked the beginning of the period of Kaptan Pasha for those who came from the palace and the Janissary Guild.[14]

This situation lasted until 1695 when Mezzomorto Hacı Huseyin Pasha reinvigorated the navy. The basic idea upon which the Ottoman Navy was based in this period was the theory of having big ships with big cannons.[15] Seeing Venetians as rivals in shipbuilding and maritime matters, the Ottoman Empire continued its activities in parallel with them. However, other Europeans and the Russians showed important technical developments in this period. The Battle of Çeşme in 1770 necessitated the modernization of their navy by the Ottomans. Ghazi Hasan Pasha of Algiers' establishment of the Imperial School of Naval Engineering/Naval Academy (Muhendishane-i Bahr-i Humayun) accelerated the navy's adaptation to technological developments and modernization.[16] Nevertheless, the Ottoman Navy was defeated by the allied navy of British, French, and Russian forces at Pylos Port in the Bay of Navarino on the Peloponnese peninsula due to the breakout of the Greek Rebellion in 1827; this defeat wreaked havoc on the Ottoman Navy.[17]

After the destruction of the fleet in Navarino in 1827, the Ottomans attempted to rebuild their navy. They ordered ships and used shipbuilding engineers from the United States of America (USA).[18] After the Russians burned the Turkish fleet in Sinop in 1853, the need to use armored ships instead of wooden boats became clear. Due to the powerful European countries' experiences in the Industrial Revolution, their incomes had increased, and they could accordingly find resources for their military expenditures to build steam and armored ships. Meanwhile, the Ottoman Empire was too far behind to be able to catch up with the Industrial Revolution. For example, Russia's military expenditure in 1855 as one of the main actors during the Crimean War was 39.8 million pounds sterling (GBP), while the Ottoman Empire's was only three million GBP. In the same year, France's military expenditure was 43.8 million GBP and Britain's was 36.5 million GBP.[19]

During the reign of Sultan Abdulaziz, the second or third strongest navy in the world and the second strongest in the Mediterranean was created in domestic and international shipyards. However, the quality was truly not that strong, because a significant number of the technical personnel on the ships were foreigners. In addition, these ships were built in different countries, so obtaining spare parts was difficult. Moreover, many foreign experts and craftsmen also worked in the shipyards, carrying out ship maintenance and repair.[20] Due to the presence of foreigners, the navy could not be used effectively. If their jobs were terminated, they were able to dismantle and take critical parts from the machines. Consequently, they had to be re-employed with higher salaries and privileges because their Turkish colleagues were unable to operate the machine.[21] Moreover, this navy brought a financial burden upon the Ottoman Empire, one that it could not afford, and the treasury went bankrupt in 1875.[22]

The negative attitudes of foreign personnel in the navy during the war, the obligation to pay foreign debts, preventing the British from destroying the navy as had occurred in Navarino, and the political concessions to Russia resulted in Sultan Abdulhamid II keeping the navy in İstanbul's Golden Horn.[23]

During the reign of Sultan Abdulhamid II, shipbuilding activities were generally carried out in Turkish shipyards. However, the ships that were built, especially until the Cretan Crisis of 1897, had been small gunboats, frigates, and torpedo boats. The most important factor behind this was the change the torpedo had created in the battlefield. In the following years, the 3,850-ton cruiser the *Hamidiye* was ordered from England in 1900 as well as the 3,485-ton cruiser the *Mecidiye* from the United States, and the 3,760-ton *Drama* was ordered from Italy in 1907.[24]

Any discussion about the presence of the Turkish Navy during the Ottoman-Italian Tripoli War and the Balkan War is not possible. Ships were almost never taken from Çanakkale to the Aegean. The loss of the Aegean islands was just as inevitable as the Balkans and was the starting point of the Aegean problems that still plague Türkiye today. Even Greek historians were astonished that the navy was not brought to the Aegean during the Balkan War, especially in light of the many trained naval officers and technical personnel whose rehabilitation works the British delegation had been carrying out for several years.[25]

According to Cemal Pasha, 24 more destroyers and 24 submarines would be needed to deter the Ottoman Empire's enemies once the peace treaty was signed at the end of the war. For this purpose, he intended to sign a contract with Germany. As a result of the meeting held in Germany between the

Ottoman delegation led by Cemal Pasha and the German delegation led by Undersecretary of the German Navy Admiral Capelle on August 29, 1917, the agreement was reached that German Naval Forces would give the Ottoman Empire 12 submarines and 12 torpedo destroyers. However, the Ottoman Empire and Germany were defeated after World War I, and Cemal Pasha, like the other Union and Progress Party leaders, fled the country. Thus, Cemal Pasha's naval project was unable to ever go beyond being a dream.[26]

In the early Republican era, Türkiye tried to fix its economy and alleviate the heavy burdens of protracted wars. At the same time, it had to take military measures to establish its security. In this respect, while priority was given to the military's land and air forces, the government invested in training navy officers and operationalizing the existing ships. During the inter-war years, Türkiye had to empower its navy in the face of Italy's revisionist policies. Due to the armament of the seas following World War I and the straits-related provisions of the Lausanne Treaty signed after the Turkish Independence War, Türkiye did not establish a navy shipyard in İstanbul.[27]

Although the impending signs of World War II and Italy's menacing policy in the Mediterranean and the Aegean Sea as well as its Abyssinian move had resulted in the acceleration of diplomatic efforts that would lead to the signing of the Montreux Convention, the time Türkiye needed to establish its navy had been eliminated. Even though Türkiye had tried to develop its German-based sea power through ship orders given to Italy and the UK before World War II, it could not take delivery of these ordered ships due to the political atmosphere of the imminent war and the unattractiveness of new ship orders in terms of Türkiye's economic situation. Thus, this period indicated the paramount importance of having a national navy and naval program.[28]

The ships purchased from Germany and the UK during World War II and the surface ships and submarines purchased from the USA after the war strengthened the navy. The Naval Undersecretariat was reconstituted as the Naval Forces Command in 1949. Having become a NATO member in 1952, Türkiye enhanced the structure, educational doctrine, opportunities, and capabilities of the Turkish Naval Forces.[29] The power of the Turkish Naval Forces increased as a consequence of modernization projects and nationalization activities in the following years. MİLGEM is the most essential of these projects and one of the most successful pieces of the Turkish defense industry.

The Route to MİLGEM

The events that aimed to make Cyprus a Greek island by eliminating all Turkish Cypriots on December 21, 1963 went down in history as Bloody Christmas and brought the possibility of a military intervention in Cyprus to Türkiye's agenda.[30] Meanwhile, the general structure of the Turkish Naval Forces had been organized entirely within the framework of NATO tasks and been equipped to achieve NATO forces' goals. The Turkish Navy consisted of ships that had been used in World War II and the Korean War and been given to Türkiye since 1949 through the US Assistance Program. The Turkish Navy had nine destroyers, ten submarines, ten patrol boats, 44 minelayers, 20 hunter boats, 42 various auxiliary ships, and nine gendarmerie boats usable in naval operations. Obviously, the main purpose in forming this navy, a third of which consisted of minelayers and search-sweepers, had been to defend the Straits against the Soviet Union.

The creation of a landing force had not been considered until that time. Unfortunately, the navy was unable to rescue the Turk cognates on the island in December 1963 because it lacked an amphibious operation doctrine on joint execution with landing troops and other forces that was capable of transferring power overseas within the force structure.[31]

However, political events in 1964 and the following years brought out the belief that national issues were unresolvable within the NATO alliance. Greek attacks against the Turks on Cyprus were ongoing, albeit generally with low sporadic intensity, up until the Peace Operation in Cyprus on July 20, 1974. These attacks had become very violent in 1964, 1967, and 1974. Within the framework of this new view toward NATO, the Turkish Naval Forces felt the need for a force capable of overseas operations when necessary. In this context, they leaned toward procuring an amphibious force with sufficient overseas transport and landing vehicles. [32]

For this purpose, establishing a Naval Society similar to the Ottoman Naval Society (Donanma-yı Osmani Muavenet-i Milliye Cemiyeti) that had been established in 1909[33] came to the fore. The "Nation Makes What Others Don't Give" campaign, was started by the *Cumhuriyet* newspaper on May 1, 1965 and gained nationwide interest. The new Naval Society contributed to the construction of ten gunboats, 12 landing craft utility (LCU) and 20 landing craft mechanized (LCM) ships as well as two destroyers to be built in the country within the five-year program organized by the Naval Forces Command under the slogan "Build your own ship".[34]

Prior to the Cyprus Peace Operation, the construction of other ships the Turkish Navy needed was additionally started, along with the formation of amphibious units. The five-year development plan spanned 1963–1967 and is accepted as having been a transition period. The priority was to construct small ships, landing ships and vehicles, auxiliary class ships, patrol ships, and gendarmerie boats.[35] The second five-year plan was arranged to cover the years 1968–1972.[36] In this context, the first ship of the project, *TCG BERK*, was laid down on January 12, 1967 and the second ship, *TCG PEYK*, on October 18, 1968; this was accomplished using national resources based on the design for the destroyer escort USS Claude Jones (DE 1033) used in the US Navy.[37]

The construction of AY class submarines started in German and Turkish shipyards, and the keels of the first two of six submarines were placed on the benches of Howaldtswerke Deutsche Werft (HDW) Shipyard in Kiel, Germany, on August 2, 1972.[38] Meanwhile, the construction of German

DOGAN Class FPB-57 fast patrol boats, one at the German Lürssen Werft Shipyard and another three at Taşkızak Shipyard, began in 1973 in order for the Turkish Navy to acquire fast patrol boats equipped with long-range Harpoon guided missiles.[39]

Sending personnel to the Naval Post-Graduate (PG) School in the USA at this time in order to pursue graduate studies in engineering was an important factor in acquiring and developing shipbuilding and repair skills. Approximately 60 officers graduated from this school between 1966 and 1977.[40] The number of graduates from this school reached 700 in 2011.[41]

One of the problems encountered after the Cyprus Peace Operation was a US-imposed arms embargo from February 5, 1975 to July 25, 1978. As a result, shipments of military supplies to Türkiye were stopped.[42] During the embargo, manufacturing spare parts that were vital for the naval units was carried out in naval shipyards and domestic industrial centers to reduce foreign dependence.[43]

The navy survived this process with relatively little damage. Although spare parts and training ammunition issues occurred, this situation created no significant problem in terms of preparing for the operation regarding the continental shelf crisis in the Aegean. The creative engineers and workers from Gölcük and Taşkızak Shipyards played an essential role in this period. The Machinery Chemistry Industry Corporation (MKEK) produced the bullets for the destroyers' 5 inch/38 caliber guns, one of the most important materials that could not be supplied due to the embargo. A branch was even established within the Supply Center Command that explored domestic supply opportunities. For example, when the pinion gear in TCG GAYRET's reduction gear mechanism would break, a local company would manufacture a new one and use it until the ship went out of service.[44]

On the other hand, this embargo revealed the drawbacks of dependency on other countries and triggered the establishment of a modern defense industry infrastructure in Türkiye. As a result of the national reaction to the embargo, the Land, Naval, and Air Force Strengthening Foundations were established.[45] In 1985, the Defense Industry Development and Support Administration was established with Law No. 3238 to develop the defense industry and modernize the Turkish Armed Forces. The Administration was restructured as the Undersecretariat for Defense Industries in 1989.[46]

Collaborations with Federal Germany during the 1980s helped Türkiye build several surface ships and submarines. The 1980s accelerated the qualitative and quantitative development the Turkish Naval Forces had started in the Republican period toward its peak. The efforts initiated to reduce the effects of the arms embargo the USA had imposed on Türkiye between 1975 and 1978 yielded results in the 1980s, and the initiatives of the naval forces at home and abroad with the aim of not being dependent on a single source for weapons and vehicles yielded effective results through modernization projects. In these years, the ships that joined the navy provided the best force structure in terms of average age and efficiency of the Republican period.[47]

The overseas individual training project was started for ships in this class in 1985 and was the most comprehensive training effort in the history of the Republic's navy. Hundreds of officers and non-commissioned officers were sent on equipment maintenance-repair and operation courses, with a large number of teachers being trained.[48]

A tank landing ship built at Taşkızak Shipyard, a fuel supply vessel built at the civil shipyard (Sedef Shipyard), and tankers built in Gölcük Shipyard were put in service during the 1990s. With these developments, the Turkish Naval Forces left behind the limited capabilities it had had in many areas in the early 1980s to become a blue-water navy in the late 1990s.[49]

The Cyprus Peace Operation and its aftermath, particularly in the late 1980s and 1990s, appeared to be the most important milestones toward MİLGEM; a project which would realize a dream for the Turkish Naval Forces and the country. For instance, the SAPAN Project that was started during the American embargo in 1975 was one of the most important projects of the Turkish Naval Forces for research and development (R&D). This project aimed to provide air defense warfare (ADW) capabilities to destroyers that had been purchased from the USA for anti-submarine warfare (ASW). Technical personnel gained self-confidence and eventually broke the belief of "we can't do it," even though the project was unable to meet its expected final goal.[50]

Moreover, the Long Horizon Project, completed in the 1990s and put in service in 2002, was a huge step in monitoring the Aegean Sea through various sensors and in transferring information to the fusion center through a linked system. Additionally, new frigates were built between 1994 and 1999 with data-sharing capabilities. Lastly, the Ship Integrated Combat Management System (GENESIS) was completed in the early 2000s and turned out to be a giant step taken toward MİLGEM. The Software Development Center (YAZGEM) was established in this context in 1994 and constituted an essential start. At the end of 1996, the Military Project Development Department (APGE) was established at the Headquarters of the Naval Forces Command. In 1998, the Research Center Command (ARMERKOM) was established in Gölcük under this department.[51]

Realizing the MİLGEM Project: "The Rain Poured Down to Hide Our Tears of Pride"

The first work on the MİLGEM project started in May 1993 with the command directive to purchase shipbuilding materials on a system/unit basis instead of purchasing materials in bulk from a single company; this enabled shipbuilding to be done with local resources.[52] The aim was to effectively use military shipyards' existing knowledge and experience, produce domestic materials at international standards, and develop sub-industry branches, as this would reduce foreign dependency and design as well as material procurement costs. The private sector would also gain experience on warships, and private Turkish shipyards would be able to build new ships by following the prototype. Accordingly, the Board of Admirals approved the MİLGEM Mission Requirements Identification (GIT) document on March 14, 1996.[53]

Within the scope of the MİLGEM project, the first Design Project Office was established in 1997 under the command of the Taşkızak Shipyard with a very limited number of personnel. Feasibility/preliminary design concept studies and technical documents regarding MİLGEM patrol and antisubmarine warfare ships were carried out by Taşkızak Shipyard Command and İstanbul Technical University. The Defense Industry Executive Committee (SSİK) decided to build the ships in 8+4 packages domestically under the responsibility of a main contractor to be determined through free competition. After the Gölcük earthquake on August 17, 1999, the Taşkızak Shipyard Command was moved to TGS Pendik Shipyard and reconstituted as İstanbul Shipyard Command. On May 18, 2002, the Undersecretariat of the Defense Industry (SSM) submitted a bid that was rescinded due to the two consortia who had responded to the call for proposals not meeting the specified requirements.[54]

After this failure, the MİLGEM Project Office (MPO) was established on March 12, 2004 to create technical specifications by bringing together the engineers, officers, and civil engineers of the Naval Forces Command under the İstanbul Shipyard Command. The MİLGEM request for bids was resent, but only one firm submitted a bid; they had no design experience and their bid was 20% over budget. As a result, this project was canceled a second time. Having Turkish engineers design the national ship under the direction of foreign designers was also considered against the spirit of this project.[55]

Due to the unexpected bid results, the need to make the design fully national emerged during the process. Thus, the Naval Forces Command took responsibility for the design and construction. Moreover, some engineering officers in the MİLGEM Project Officer Directorship said, "We can design and integrate MİLGEM in this office. We only need trust, additional engineers, and timely procurement of material and services. Therefore, the result will eventually be our National Ship"; this affected the above decision [49]. In this procurement model, the design and performance were entirely under the Naval Forces Command's responsibility, alongside various contractors such as the Aselsan/Havelsan Joint Venture, War Technologies Engineering and Trade (STM A.Ş.), Motor Turbine Industry and Trade (MTU Türkiye) Türk Loydu Foundation, and İstanbul Technical University. Dozens of large and small domestic defense and industrial companies were also encouraged to participate in the project.[56]

A warship's most important design criterion is its ability to survive on the battlefield. This criterion had been met in the first half of the 20th century with thick/protective armor-covered ships and long-range artillery. However, today, this is achieved with track management measures, long-range sensors, and weapons that will avoid timely enemy detection thus providing first strike opportunity. In this context, the MİLGEM Project Office Directorate has taken into consideration modern measures for minimizing ship signatures (e.g., underwater, cross-sectional radar area, magnetic, and infrared traces) throughout the entire design process.[57]

In addition, Lloyd's Register's Rules and Regulations for the Classification of Naval Ships (known as Türk Loydu Naval Ship Rules in Türkiye) were published for the first time in Türkiye and implemented in the design. In this context, strength and vibration analyses of the ship's construction were made using structural dimensioning calculations. These studies contributed significantly to the development of the Türk Loydu rules.[58] With this project, an additional non-contact underwater explosion strength analysis of a battleship against torpedo and mine weapons was made for the first time in Türkiye. The hull elements were designed based on these analyses.[59]

The principle of the project was to use national resources to the maximum extent possible. As a result of the collaboration between ITU and the MİLGEM Project Office, form and hydrodynamic-design analyses were made, with the hull form and elements such as rudder, shaft, shaft bearing, and yaw corrector blades being locally produced. The construction of *TCG HEYBELİADA* started on January 22, 2007 at İstanbul Shipyard Command.[60]

Having been involved in the project since the beginning, retired Captain (N) Kerem Orçun Yüksekdağ, Naval Engineer, MSc described what he felt on that day as follows:

> The pride we felt, the tears of joy we shed are still in my mind with the same clarity as that day, and I will never forget it for the rest of my life. Our long-sleeved white uniforms on, the pouring rain in İstanbul, the excitement I felt as if my heart were going to burst. My prayers for the ship for which I'd made all the calculations for the launching to the blue homeland without any problem... Stealing one of the yellow raincoats given only to the guests in the protocol, hiding behind an electric box with a stopwatch, my shoes as soaked as a lake, the endless ceremonial speeches, and finally that moment... The *HEYBELİADA*, moving like a delicate bride, accelerated as she approached the sea. When I turned my head to the left, the whole office personnel were running toward me, hugging and crying. I experienced such an intensity of emotion only at the birth of my children and no other time...[61]

The MİLGEM Project Office carried out intensive system-system and system-ship integration design activities for *TCG HEYBELİADA*'s construction. Integration design is a detailed placement process conducted on a three-dimensional model in a way that gathers product/system design information from sub-contractors with regard to boat, machinery, electrical, and warfare system/equipment to ensure they will meet both the systems' design conditions as well as the technical, functional, ergonomic, and maritime integration conditions.[62]

In fact, the entire systems of the ship were not created domestically. The ship's main propulsion system was procured from a company called MTU Aero Engines in Germany. Nevertheless, assembling such an enormous system on the ship would be carried out by Turkish engineers and workers in the Turkish shipyard for the first time ever. This situation meant gaining broad experience. Retired Captain (N) Zafer Elçin, Mechanical Engineer MSc, one of the main propulsion engineers within the MİLGEM Project Office, stated the following about this issue:

> The main diesel engines, shafts, and propellers were brought aboard and mounted in the dry dock after the gas turbines and auxiliary equipment. Despite the incredible size of the reduction gear, which to me is the heart of the main drive system, the assembly procedure also required a lot of precision. The reduction gear consists of three parts and weighs approximately 150 tons in total. It was to be assembled with tolerances in hundredths of a millimeter after the ship was relaunched.

> …While checking the accuracy of the alignment process, the load bearings on one of the two propeller shafts on the ship was not within the desired values. The contractor and shipyard staff worked for several months to achieve ideal payloads. However, the result was never at the expected level. Eventually, the contractors brought a 70- or 80-year-old specialist from Germany to the shipyard to help us solve the problem. After working with us for a day, the German specialist locked himself in a container next to the ship. He did some work inside for two days. He then gave us the new calculations to be applied to the system. According to the new measurements at the end of the alignment process, the standards for the load bearings had reached the desired level. When I spoke to the German expert, I asked him which computer program he used. The German told me he did the calculations by hand, which is why it took so long. I realized once again that we still have a lot to learn.[63]

In addition, the system-equipment product development activities that are able to meet the technical and integration conditions of the ship are carried out together with the sub-product contractors in the integration design. Systems and equipment developed within the scope of the nationally designed MİLGEM Project and produced for the first time in Türkiye include the use of shipbuilding sheet metal with high yield strength, rudder systems, shock resistant hulls, system/equipment shock absorbers, freshwater generation system, electrical distribution unit systems, air intake/exhaust systems, compressed air systems, nationally designed fire suppression system, ship anchor-windlass systems, telescopic bollards, vehicle davits-ship vehicles, ship-rated furniture, service area equipment, helicopter hangar door, torpedo sleeve, aft vehicle covers, starboard piers, and window glass and casing.[64]

This process ensured that the steel hull model had been created, the circuit designs had been made, the system/equipment conflict analyses had been performed, and assembly/workmanship drawings had been produced for all steel and equipment operations based on quality workmanship. The integration design required intensive engineering effort from the MİLGEM Project

Office until the ship was put into service. In order to transfer the significant success and design capability of the MİLGEM project to the subsequent Naval Forces projects, the MİLGEM Project Office produced approximately 12,000 pictures and documents for *TCG HEYBELİADA*'s construction; their name was also changed to the Design Project Office Directorate (DPO) in 2010.[65]

The *TCG HEYBELİADA* (F-511) was put into service on September 27, 2011. The MİLGEM is one of the rare projects among its counterparts to have been completed within the stipulated time and to have a 65% domestic production rate. One of the important factors behind this success is that the officers who had graduated from the Naval War Academy's engineering program had been assigned to the engineer class and taken part in the project after serving on ships for a certain period. This policy helped them build a common understanding with the end-users.[66]

Through the MİLGEM Project, the design and integration of a complex surface warship with high standards and combat power in accordance with today's technology were carried out for the first time in Türkiye using national resources with the support of domestic industries. Moreover, this project paved the way for the nationalization of several critical systems. Lastly, Türkiye became one of the countries with national combat ship design and construction capabilities due to this project.[67]

The sonar devices used in the MİLGEM Project were also developed domestically. Lieutenant Commander Ceyhun İlgüy, Acoustic Engineer, MSc, was one of three engineers working on the national sonar system within the Research Center Command and cited his experiences on this subject as follows:

> We started this project as three people who wholeheartedly believed what we needed were dedication and high motivation. However, failure was more likely, and no one even knew how to start it. Project management had the features and costs of the systems collected from foreign companies as a 'backup plan'. If we had not succeeded, one of those systems would have been purchased.[68]

Working together under this high stress, the acoustics and software development group initially did not understand each other. However, they found a high level of harmony afterwards. Lieutenant Colonel İlgüy stated that he has worked in many domestic and foreign companies since this project but never witnessed such cohesion and work discipline again. Detecting a target 21,000 yards away meant they had successfully completed the acceptance test. İlgüy further expressed his feelings as follows, which are valuable for understanding the mindset of the project team:

> I experienced that day thanks to those who believed in this project. When I think about it again, in the beginning, there was no 'me', but there was 'we'. It looks like a fairytale today. Many materials were procured from

abroad, but everything, even the power supplies, was developed in Türkiye. It was unnecessary as there were so many finished products, but it was an obsession. Put simply, everything needed to be domestic.[69]

At the same time, the project provided self-confidence to National Defense Industry companies. The construction of the project's four ships was completed with the integration of the new national systems into the design of the ADA Class Corvettes as they became ready. Another common feature of domestically produced ships is that they emphasize the homeland in their emblems, which they carry from their first to last day of service. For example, *TCG HEYBELİADA* has the motto "Our Strength is from the Homeland," *TCG BÜYÜKADA* has the motto "Homeland is at Sea," *TCG BURGAZADA* has the motto "Fortress of the Blue Homeland," and *TCG KINALIADA* has the motto "Devoted to the Blue Homeland".[70]

One of the main determinants of the project's success was that the distinguished Naval War Academy graduates later received postgraduate education at elite universities in the country and abroad. However, the education dimension of the project has its roots dating back to the 1990s. As a result of the restructuring of the Naval Academy curriculum at that time and the dedication and enthusiasm of the officers for self-development, many officers had the opportunity to receive postgraduate education at other universities without the need for additional engineering-related courses.[71] Lastly and more importantly, the decision to assign the best officers to the engineering branch rather than the critical operational posts proved the visionary outlook of navy leadership.

Conclusion

Once the masters of the Black Sea and the Mediterranean, Turks lost their superiority through various naval battles starting with Lepanto in 1571. The delay in transitioning from rowing to sailing caused Turks the absolute loss of Mediterranean sovereignty. After each defeat, ships were built quickly; however, filling the places of the trained personnel and valuable captains who had been lost took a longer amount of time. The resulting inability to produce ships had constantly brought foreign dependency. In addition, weakness had been found in training competent seafarers, especially technical personnel.

In the first years of the Republic, the navy consisted of a small number of wounded and old ships. Moreover, the Golden Horn as the epicenter of the navy where the most important shipyard had been located was unusable due to being in the demilitarized zone in accordance with the Lausanne Treaty. Therefore, in order to remove this obstacle, the new Republic considered establishing a new shipyard and base in Gölcük as its primary goal.

After World War II and especially after becoming a NATO member in 1952, Türkiye structured its navy to serve the common interests of the Alliance. The negative impacts of this policy came to light during the Cyprus crisis that started in 1963, with the efforts of having a navy capable of overseas operations gaining momentum. Between 1964 and 1974, the supply and construction of landing ships in particular were accelerated. The arms embargo the USA implemented after the 1974 Cyprus Peace Operation revealed the need to eliminate foreign dependency. The need to make materials domestically triggered the foundations for innovative work in the navy.

New ships were purchased from Germany in place of the old vessels the USA had given; some of the purchased ships were built in shipyards in Türkiye, but this was assembly, not manufacturing. The number of engineers who have been sent to the Naval Post-Graduate School since 1966 has increased, and gradually those officers have assumed decision-making positions. They accelerated the innovation efforts combined with the experience gained during the assembly process. Additionally, the embargo period turned into a positive experience that helped them develop the self-confidence of "we can do it."

The officers were selected from the cream of the navy and became highly trained engineers. Their dominant positions on the Board of Admirals and superior headquarters were vital to initiating major naval projects, including MİLGEM. This visionary perspective not only managed to acquire political support from all the governments of the period but also attracted the support of civil defense industry organizations and universities. As a result, the projects carried out were based on the participation of all stakeholders. The opportunities and capabilities of civil institutions and organizations were fully utilized, and a great synergy was created by including many sub-contractors in the projects. These collaborations are considered as the most important factor in turning this snowball into an avalanche.

Notes

1 İdris Bostan & Salih Özbaran, "Giriş", in (eds.) İdris Bostan & Salih Özbaran, *Başlangıçtan XVII. Yüzyılın Sonuna Kadar Türk Denizcilik Tarihi I*, İstanbul, Deniz Basımevi, 2009, pp. 11–13. İlhami Durmuş, *İskitler*, Hacettepe Üniversitesi Sosyal Bilimler Enstitüsü, Ankara, 1992, p. 1. Akdes Nimet Kurat, *IV-XVIII. Yüzyıllarda Karadeniz Kuzeyindeki Türk Kavimleri ve Devletleri*, Türk Tarih Kurumu Yayınları, Ankara, 1972, p. 291. İbrahim Kafesoğlu, *Harzemşahlar Devleti Tarihi*, Türk Tarih Kurumu Yayınları, Ankara, 1984, p. 258.
2 Hayati Tezel, *Anadolu Türklerinin Deniz Tarihi*, Deniz Basımevi, İstanbul, 1973, p. 3.
3 Mustafa Daş, "Türklerin Bizans ve Venedik'le Denizdeki İlişki ve Mücadeleleri (XI-XIV. Yüzyıllar), in (eds.) İdris Bostan & Salih Özbaran, Başlangıçtan XVII. Yüzyılın Sonuna Kadar Türk Denizcilik Tarihi I, İstanbul, Deniz Basımevi Müdürlüğü, 2009, pp. 49–50. Mücteba İlgürel, "Osmanlı Denizciliğinin İlk Devirleri", *Belleten*, LXV:243, 2001, pp. 637–653.
4 Anna Comnena, *The Alexiad*, (transl.) Elizabeth A. S. Dawes, In parentheses Publications, Ontario 2000, p. 129.
5 Halil Özsaraç, *Donanmanın Tarihsel Serüveni*, Doruk Yayınları, İstanbul, 2020, p. 11.
6 Halil İnalcık, "The Rise of the Turcoman Maritime Principalities in Anatolia, Byzantium and the Crusades", *The Middle East and the Balkans under the Ottoman Empire Essays on Economy and Society*, Bloomington, 1993, pp. 309–341.

90   *Kemal Eker*

7 Nuri Alacalı, Nurcan Bal & Figen Atabey, *Türk Deniz Kuvvetleri Bin Yılın Güncesinden Seçmeler*, Deniz Basımevi, İstanbul, 2009, p. 9.
8 İdris Bostan, *Beylikten İmparatorluğa Osmanlı Denizciliği*, Kitap Yayınevi, İstanbul, 2007, p. 15. Kate Fleet, "Early Turkish Naval Activities", *Oriente Moderno*, 81:1, 2001, p. 135.
9 İdris Bostan, *Osmanlı Bahriye Teşkilatı: XVII. Yüzyılda Tersâne-i Âmire*, Türk Tarih Kurumu Yayınları, Ankara, 1992, pp. 1–14.
10 Piri Reis, *Kitabı Bahriye*, Türk Tarih Kurumu Yayınları, Ankara, 2002.
11 İdris Bostan, "Kadırga'dan Kalyon'a XVII. Yüzyılın İkinci Yarısında Osmanlı Gemi Teknolojisinin Değişimi", *Osmanlı Araştırmaları*, 24:24, 2004, pp. 65–86.
12 Fernand Braudel, *Akdeniz ve Akdeniz Dünyası*, (trans.) Mehmet Ali Kılıçbay, Eren Yayıncılık, İstanbul, 1990, p. 328.
13 İdris Bostan, "Malta Kuşatmasından Tunus'un Fethine", in (eds.) İdris Bostan & Salih Özbaran, *Başlangıçtan XVII. Yüzyılın Sonuna Kadar Türk Denizcilik Tarihi I*, İstanbul, Deniz Basımevi, 2009, pp. 195–196. Halil İnalcık, "Barbaros'tan İnebahtı (Leponto)'ya Akdeniz", in (ed.) Bülent Arı, *Türk Denizcilik Tarihi*, T.C. Başbakanlık Denizcilik Müsteşarlığı Yayınları, Ankara 2002, p. 142.
14 İsmail Hakkı Uzunçarşılı, *Osmanlı Devleti'nin Merkez ve Bahriye Teşkilatı*, Türk Tarih Kurumu Yayınları, Ankara 1984, pp. 495–496. İbrahim Akkaya & Fahri Ayanoğlu, *Osmanlı İmparatorluğu'ndan Günümüze Denizlerimizin Amirleri, Derya Kaptanları, Bahriye Nazırları ve Deniz Kuvvetleri Komutanları*, Deniz Basımevi, İstanbul, 2009, p. 39.
15 Afif Büyüktuğrul, *Osmanlı Deniz Harp Tarihi ve Cumhuriyet Donanması* II, Deniz Basımevi, İstanbul, 1970, p. 200. Saffet, *Mezemorta Hüseyin Paşa*, (trans.) Yavuz Senemoğlu, Deniz Basımevi, İstanbul, 1994.
16 Ali İhsan Gencer, *Bahriye'de Yapılan Islahat Hareketleri ve Bahriye Nezareti'nin Kuruluşu (1789–1867)*, Türk Tarih Kurumu Yayınları, Ankara, 2001, pp. 83–84. Tezel, *Anadolu Türklerinin Deniz Tarihi*, pp. 327–329, 349–351, 400–403. Uzunçarşılı, *Osmanlı Devleti'nin Merkez ve Bahriye Teşkilatı*, pp. 507–511.
17 Ali Fuat Örenç, "1827 Navarin Deniz Savaşı ve Osmanlı Donanması", *Tarih Dergisi*, 46, İstanbul, 2009, pp. 37–84. Tezel, *Anadolu Türklerinin Deniz Tarihi*, pp. 400–403.
18 Çağrı Erhan, *Türk-Amerikan İlişkilerinin Tarihsel Kökenleri*, İmge Kitabevi Yayınları, İstanbul, 2015, p. 132.
19 Paul Kennedy, *Büyük Güçlerin Yükseliş ve Çöküşleri*, (trans.) Birtane Karanakçı, Türkiye İş Bankası Kültür Yayınları, İstanbul, 2017, p. 207.
20 Bernd Langensiepen & Ahmet Güleryüz, *The Ottoman Steam Navy 1828–1923*, Naval Institute Press, Annapolis, 1995.
21 Celalettin Yavuz, "Sultan Abdülaziz Donanması: Yelkenli Teknelerden Buhar Makineli Gemilere Geçiş, Bitmeyen Reform İhtiyaçları", *XIII. Türk Tarih Kongresi*, Türk Tarih Kurumu, Ankara, 2002, pp. 1816–1817. Henry F. Woods, *Türkiye Anıları*, (trans.) Fahri Çoker, Milliyet Yayınları, İstanbul, 1976, p. 40.
22 Enver Ziya Karal, *Osmanlı Tarihi VII*, Türk Tarih Kurumu Yayınları, Ankara, 2011, p. 237. Yavuz, "Sultan Abdülaziz Donanması-Yelkenli Teknelerden Buhar Makineli Gemilere Geçiş, Bitmeyen Reform İhtiyaçları" p. 1817.
23 Kaori Komatsu, "Financial Problems of the Navy During the Reign of Abdülhamid II", *Oriente Moderno*, 20:81, 2001, p. 209. Afif Büyüktuğrul, *Osmanlı Deniz Harp Tarihi ve Cumhuriyet Donanması III*, Deniz Basımevi, İstanbul, 1970, p. 166.
24 Yavuz, "Sultan Abdülaziz Donanması: Yelkenli Teknelerden Buhar Makineli Gemilere Geçiş, Bitmeyen Reform İhtiyaçları", pp. 1827–1832.
25 O. Ramis & Y. Teofanidis, *Türk ve Yunan Deniz Harbi Hatıratı ve 1909–1913 Yunan Bahri Tarihi*, Büyük Erkânıharbiye XII. Deniz Şubesi, İstanbul, 1930, p. 59.
26 Celalettin Yavuz, "Cemal Paşa'nın Almanya Gezisi: Bir Davetin Perde Arkası", *Atatürk Yolu*, 5:19, 1997, pp. 335–346.

27 Raşit Metel, *Atatürk ve Donanma*, Deniz Basımevi, İstanbul, 1966.
28 Serhat Güvenç & Dilek Barlas, "Atatürk's Navy: Determinants of Turkish Naval Policy, 1923–38." *Journal of Strategic Studies*, 26:1, 2008, pp. 23–26. Afif Büyüktuğrul, *Büyük Atamız ve Türk Denizciliği*, Deniz Basımevi, İstanbul, 2006. Fahri Çoker, *Bahriyemizin Yakın Tarihinden Kesitler*, Deniz Kuvvetleri Basımevi, Ankara, 1994. Yüksel Öcal, *Kürek ve Yelken Döneminden Günümüze Türk Bahriyesi*, Deniz Basımevi, İstanbul, 2008.
29 Afif Büyüktuğrul, *Cumhuriyet Donanması (1923–1960)*, Deniz Basımevi, İstanbul, 1967, pp. 104–106.
30 Abdulhaluk Çay, *Kıbrıs'ta Kanlı Noel*, Türk Kültürünü Araştırma Enstitüsü Yayınları, Ankara, 1989. Harry Scott Gibbons, *Kıbrıs'ta Soykırım*, (trans.) Erol Fehim, Özyurt Matbaası, Ankara, 2003.
31 Cem Gürdeniz, *Hedefteki Donanma*, Kırmızı Kedi Yayınevi, İstanbul, 2013, pp. 115–116. Y. Necdet Çelik & Erdoğan Yüceliş, *Cumhuriyet Donanması 1923–2005*, Deniz Basımevi, İstanbul, 2005.
32 Figen Atabey (ed.), *Cumhuriyet Dönemi Türk Deniz Kuvvetleri*, Deniz Kuvvetleri Komutanlığı Karargâh Basımevi, Ankara, 2002, pp. 59–60. Nihat Erim, *Bildiğim ve Gördüğüm Ölçüler İçinde Kıbrıs*, Ajans-Türk Matbaacılık, Ankara, 1975. Pierre Oberling, *Bellapais'e Giden Yol: Kıbrıs Türklerinin Kuzey Kıbrıs'a Göçü*, (trans.) Mehmet Erdoğan, Genelkurmay Basımevi, Ankara, 1988. Sevin Toluner, *Kıbrıs Uyuşmazlığı ve Milletlerarası Hukuk*, İstanbul Üniversitesi Yayınları, İstanbul, 1977.
33 Selahattin Özçelik, *Donanma-yı Osmani Muavenet-i Milliye Cemiyeti*, TTK Basımevi, Ankara, 2000.
34 Çoker, *Bahriyemizin Yakın Tarihinden Kesitler*, pp. 25–27.
35 Sina Özdoğancı, "Türk Deniz Kuvvetlerinde Gemi İnşa Faaliyetleri ve Hamleleri", *Deniz Kuvvetleri Dergisi*, 74:460, 1968, p. 7.
36 Sait Arif Terzioğlu, "Şilep Yapmak Muhrip Yapmaktan Kolaydır", *Derya Denizcilik Dergisi*, 56, 1972, p. 10.
37 Gürdeniz, *Hedefteki Donanma*, p. 135.
38 Denizaltı Filosu Komutanlığı, *Sessiz ve Derinden: Barışın Koruyucusu Geleceğin Güvencesi Denizaltılarımız*, Deniz Kuvvetleri Komutanlığı Seyir Hidrografi ve Oşinografi Dairesi Başkanlığı Yayını, İstanbul, 2007, p. 129.
39 Hücumbot Filosu Komutanlığı, *Hücumbotlar: Rüzgârla Yarışanlar*, Deniz Basımevi Müdürlüğü, İstanbul, 2008, p. 131.
40 Özden Örnek, *MİLGEM'in Öyküsü*, Kırmızı Kedi Yayınevi, İstanbul, 2016, pp. 21–22.
41 Gürdeniz, *Hedefteki Donanma*, p. 138.
42 Örnek, *MİLGEM'in Öyküsü*, p. 31.
43 Atabey, *Cumhuriyet Dönemi Türk Deniz Kuvvetleri*, p. 64.
44 Örnek, *MİLGEM'in Öyküsü*, p. 33.
45 "Tarihçe", Türk Silahlı Kuvvetlerini Güçlendirme Vakfı, https://www.tskgv.org.tr/tr/hakkimizda/tarihce, accessed on 13.03.2022.
46 "Hakkımızda", Savunma Sanayi Başkanlığı, https://www.ssb.gov.tr/WebSite/ContentList.aspx?PageID=39, accessed on 13.03.2022.
47 Atabey, *Cumhuriyet Dönemi Türk Deniz Kuvvetleri*, p. 64.
48 Gürdeniz, *Hedefteki Donanma*, pp. 191–195.
49 Atabey, *Cumhuriyet Dönemi Türk Deniz Kuvvetleri*, p. 64.
50 Örnek, *MİLGEM'in Öyküsü*, pp. 38–44.
51 Örnek, *MİLGEM'in Öyküsü*, pp. 58–81.
52 Örnek, *MİLGEM'in Öyküsü*, p. 82.
53 Timur Diler, Fatih Piren, Fuat Çelik & İ. Mehmet Doğutepe, "MİLGEM Projesi'nin Tarihsel Süreci", *Mavi Vatan'dan Açık Denizlere*, 8, 2021, p. 84. Örnek, *MİLGEM'in Öyküsü*, p. 85.

54 Diler, et al. "MİLGEM Projesi'nin Tarihsel Süreci", p. 84. Örnek, *MİLGEM'in Öyküsü*, pp. 85–92.
55 Diler, et al. "MİLGEM Projesi'nin Tarihsel Süreci", p. 84. Örnek, *MİLGEM'in Öyküsü*, pp. 85–104.
56 Diler, et al. "MİLGEM Projesi'nin Tarihsel Süreci", p. 85. Kadir A. Demir, Ebru Caymaz & Fahri Erenel, "Defense Industry Clusters in Turkey", *Journal of Defense Resources Management*, 7:1, 2016, p. 9.
57 Diler, et al. "MİLGEM Projesi'nin Tarihsel Süreci", p. 85. Börteçin Ege, "MİLGEM", *Bilim ve Teknik*, 48:563, 2014, pp. 56–57.
58 Örnek, *MİLGEM'in Öyküsü*, pp. 164–166.
59 Diler, et al. "MİLGEM Projesi'nin Tarihsel Süreci", p. 86.
60 Örnek, *MİLGEM'in Öyküsü*, pp. 185–187. Diler, et al. "MİLGEM Projesi'nin Tarihsel Süreci", p. 86. "İstanbul Naval Shipyard: A Major Player in Turkish Naval Projects", *Defence Turkey*, https://www.defenceturkey.com/en/content/istanbul-naval-shipyard-a-major-player-in-turkish-naval-projects-298, accessed on 13.03.2022.
61 Interview with Retired Captain (N) Kerem Orçun Yüksekdağ on December 03, 2021.
62 "TCG Heybeliada-(MİLGEM-Milli Gemi-National Ship) Patrol and Anti-Submarine Warfare Ship" https://www.globalsecurity.org/military/world/europe/tcg-milgem.htm, accessed on 13.03.2022. Diler, et al. "MİLGEM Projesi'nin Tarihsel Süreci", pp. 86–87.
63 Interview with Retired Captain (N) Zafer Elçin on December 05, 2021.
64 "MİLGEM Project", https://www.stm.com.tr/en/our-solutions/naval-engineering/milgem-project, accessed on 13.03.2022. Diler, et al. "MİLGEM Projesi'nin Tarihsel Süreci", p. 87.
65 Tresno Wicaksono & Anak Agung Banyu Perwita, "The Military Industrial Complex in a Developing Country: Lessons from the Republic of Turkey", *Jurnal Hubungan Internasional*, 9(1), 2020, p. 59. Diler, et al. "MİLGEM Projesi'nin Tarihsel Süreci", p. 87.
66 Diler, et al. "MİLGEM Projesi'nin Tarihsel Süreci", p. 87.
67 Diler, et al. "MİLGEM Projesi'nin Tarihsel Süreci", pp. 87–88. Ege, "MİLGEM", p. 57.
68 Interview with ex-Lieutenant Commander Ceyhun İlgüy on December 11, 2021.
69 Interview with ex-Lieutenant Commander Ceyhun İlgüy on December 11, 2021.
70 "ADA Class (MİLGEM)", Bosphorus Naval News, https://turkishnavy.net/ada-class-milgem, accessed on 13.03.2022. Diler, et al. "MİLGEM Projesi'nin Tarihsel Süreci", p. 88.
71 "History of Turkish Naval Academy", https://dho.msu.edu.tr/sayfalar/00_Anasayfa/01_Sabitler/tarihce/tarihce.html, accessed on 13.03.2022.

References

"ADA Class (MİLGEM)", Bosphorus Naval News, https://turkishnavy.net/ada-class-milgem.
Akkaya, İbrahim & Fahri Ayanoğlu, *Osmanlı İmparatorluğu'ndan Günümüze Denizlerimizin Amirleri, Derya Kaptanları, Bahriye Nazırları ve Deniz Kuvvetleri Komutanları*, Deniz Basımevi, İstanbul, 2009.
Alacalı, Nuri, Nurcan Bal & Figen Atabey, *Türk Deniz Kuvvetleri Bin Yılın Güncesinden Seçmeler*, Deniz Basımevi, İstanbul, 2009.
Atabey, Figen (ed.), *Cumhuriyet Dönemi Türk Deniz Kuvvetleri*, Deniz Kuvvetleri Komutanlığı Karargâh Basımevi, Ankara, 2002.
Bostan, İdris, *Osmanlı Bahriye Teşkilatı: XVII. Yüzyılda Tersâne-i Âmire*, Türk Tarih Kurumu Yayınları, Ankara, 1992.

Bostan, İdris, "Kadırga'dan Kalyon'a XVII. Yüzyılın İkinci Yarısında Osmanlı Gemi Teknolojisinin Değişimi", *Osmanlı Araştırmaları*, 24:24, 2004, pp. 65–86.

Bostan, İdris, *Beylikten İmparatorluğa Osmanlı Denizciliği*, Kitap Yayınevi, İstanbul, 2007.

Bostan, İdris, "Malta Kuşatmasından Tunus'un Fethine", in (eds.) İdris Bostan & Salih Özbaran, *Başlangıçtan XVII. Yüzyılın Sonuna Kadar Türk Denizcilik Tarihi I*, Deniz Basımevi, İstanbul, 2009, pp. 185–197.

Bostan, İdris & Salih Özbaran, "Giriş", in (eds.) İdris Bostan & Salih Özbaran, *Başlangıçtan XVII. Yüzyılın Sonuna Kadar Türk Denizcilik Tarihi I*, İstanbul, Deniz Basımevi, 2009, pp. 11–16.

Braudel, Fernand, *Akdeniz ve Akdeniz Dünyası*, (trans.) Mehmet Ali Kılıçbay, Eren Yayıncılık, İstanbul, 1990.

Büyüktuğrul, Afif, *Cumhuriyet Donanması (1923–1960)*, Deniz Basımevi, İstanbul, 1967.

Büyüktuğrul, Afif, *Osmanlı Deniz Harp Tarihi ve Cumhuriyet Donanması Vols. II-III*, Deniz Basımevi, İstanbul, 1970.

Büyüktuğrul, Afif, *Büyük Atamız ve Türk Denizciliği*, Deniz Basımevi, İstanbul, 2006.

Çay, Abdulhaluk, *Kıbrıs'ta Kanlı Noel*, Türk Kültürünü Araştırma Enstitüsü Yayınları, Ankara, 1989.

Çelik, Y. Necdet and Yüceliş, Erdoğan, *Cumhuriyet Donanması 1923–2005*, Deniz Basımevi, İstanbul, 2005.

Çoker, Fahri, *Bahriyemizin Yakın Tarihinden Kesitler*, Deniz Kuvvetleri Komutanlığı Basımevi, Ankara, 1994.

Comnena, Anna, *The Alexiad*, (transl.) Elizabeth A. S. Dawes, In parentheses Publications, Cambridge, Ontario 2000.

Daş, Mustafa, "Türklerin Bizans ve Venedik'le Denizdeki İlişki ve Mücadeleleri (XI-XIV. Yüzyıllar)", in (eds.) İdris Bostan & Salih Özbaran, *Başlangıçtan XVII. Yüzyılın Sonuna Kadar Türk Denizcilik Tarihi I*, İstanbul, Deniz Basımevi Müdürlüğü, 2009, pp. 49–59.

Demir, Kadir A., Ebru Caymaz & Fahri Erenel, "Defense Industry Clusters in Turkey", *Journal of Defense Resources Management*, 7:1, 2016, pp. 7–20.

Denizaltı Filosu Komutanlığı, *Sessiz ve Derinden: Barışın Koruyucusu Geleceğin Güvencesi Denizaltılarımız*, Deniz Kuvvetleri Komutanlığı Seyir Hidrografi ve Oşinografi Dairesi Başkanlığı Yayını, İstanbul, 2007.

Diler, Timur, et al, "MİLGEM Projesi'nin Tarihsel Süreci", *Mavi Vatan'dan Açık Denizlere*, 8, 2021, pp. 70–92.

Durmuş, İlhami, *İskitler*, Hacettepe Üniversitesi Sosyal Bilimler Enstitüsü, Ankara, 1992.

Ege, Börteçin, "MİLGEM", *Bilim ve Teknik*, 48:563, 2014, pp. 56–57.

Erhan, Çağrı, *Türk-Amerikan İlişkilerinin Tarihsel Kökenleri*, İmge Kitabevi Yayınları, İstanbul, 2015.

Erim, Nihat, *Bildiğim ve Gördüğüm Ölçüler İçinde Kıbrıs*, Ajans-Türk Matbaacılık, Ankara, 1975.

Fleet, Kate, "Early Turkish Naval Activities", *Oriente Moderno*, 81:1, 2001, pp. 129–138.

Gencer, Ali İhsan, *Bahriye'de Yapılan Islahat Hareketleri ve Bahriye Nezareti'nin Kuruluşu (1789–1867)*, Türk Tarih Kurumu Yayınları, Ankara, 2001.

Gibbons, Harry Scott, *Kıbrıs'ta Soykırım*, (trans.) Erol Fehim, Özyurt Matbaası, Ankara, 2003.

Gürdeniz, Cem, *Hedefteki Donanma*, Kırmızı Kedi Yayınevi, İstanbul, 2013.

Güvenç, Serhat & Dilek Barlas, "Atatürk's Navy: Determinants of Turkish Naval Policy, 1923–38." *Journal of Strategic Studies*, 26:1 2008, pp. 1–35.

"Hakkımızda", Savunma Sanayi Başkanlığı, https://www.ssb.gov.tr/WebSite/ContentList.aspx?PageID=39.

"History of Turkish Naval Academy", https://dho.msu.edu.tr/sayfalar/00_Anasayfa/01_Sabitler/tarihce/tarihce.html.

Hücumbot Filosu Komutanlığı, *Hücumbotlar: Rüzgârla Yarışanlar*, Deniz Basımevi Müdürlüğü, İstanbul, 2008.

İlgürel, Mücteba, "Osmanlı Denizciliğinin İlk Devirleri", *Belleten*, LXV:243, 2001, pp. 637–653.

İnalcık, Halil, "The Rise of the Turcoman Maritime Principalities in Anatolia, Byzantium and the Crusades", in (ed.) Halil İnalcık, *The Middle East and the Balkans under the Ottoman Empire Essays on Economy and Society*, Indiana University Turkish Studies, Bloornington, 1993, pp. 309–341.

İnalcık, Halil, "Barbaros'tan İnebahtı (Leponto)'ya Akdeniz", in (ed.) Bülent Arı, *Türk Denizcilik Tarihi*, T.C. Başbakanlık Denizcilik Müsteşarlığı Yayınları, Ankara 2002, pp. 141–144.

Interview with ex Lieutenant Commander Ceyhun İlgüy on December 11, 2021.

Interview with Retired Captain (N) Kerem Orçun Yüksekdağ on December 03, 2021.

Interview with Retired Captain (N) Zafer Elçin on December 05, 2021.

"İstanbul Naval Shipyard: A Major Player in Turkish Naval Projects", *Defence Turkey*, https://www.defenceturkey.com/en/content/istanbul-naval-shipyard-a-major-player-in-turkish-naval-projects-298.

Kafesoğlu, İbrahim, *Harzemşahlar Devleti Tarihi*, Türk Tarih Kurumu Yayınları, Ankara, 1984.

Karal, Enver Ziya, *Osmanlı Tarihi VII*, Türk Tarih Kurumu Yayınları, Ankara, 2011.

Kennedy, Paul, *Büyük Güçlerin Yükseliş ve Çöküşleri*, (trans.) Birtane Karanakçı, Türkiye İş Bankası Kültür Yayınları, İstanbul, 2017.

Komatsu, Kaori, "Financial Problems of the Navy During the Reign of Abdülhamid II", *Oriente Moderno*, 20:81, 2001, pp. 209–219.

Kurat, Akdes Nimet, *IV-XVIII. Yüzyıllarda Karadeniz Kuzeyindeki Türk Kavimleri ve Devletleri*, Türk Tarih Kurumu Yayınları, Ankara, 1972.

Langensiepen, Bernd & Ahmet Güleryüz, *The Ottoman Steam Navy 1828–1923*, Naval Institute Press, Annapolis, 1995.

Metel, Raşit, *Atatürk ve Donanma*, Deniz Basımevi, İstanbul, 1966.

"MİLGEM Project", https://www.stm.com.tr/en/our-solutions/naval-engineering/milgem-project.

Oberling, Pierre, *Bellapais'e Giden Yol: Kıbrıs Türklerinin Kuzey Kıbrıs'a Göçü*, (trans.) Mehmet Erdoğan, Genelkurmay Basımevi, Ankara, 1988.

Öcal, Yüksel, *Kürek ve Yelken Döneminden Günümüze Türk Bahriyesi*, Deniz Basımevi, İstanbul, 2008.

Örenç, Ali Fuat, "1827 Navarin Deniz Savaşı ve Osmanlı Donanması", *Tarih Dergisi*, 46, İstanbul, 2009, pp. 37–84.

Örnek, Özden, MİLGEM'in Öyküsü, Kırmızı Kedi Yayınevi, İstanbul, 2016.

Özçelik, Selahattin, *Donanma-yı Osmani Muavenet-i Milliye Cemiyeti*, Türk Tarih Kurumu Basımevi, Ankara, 2000.

Özdoğancı, Sina, "Türk Deniz Kuvvetlerinde Gemi İnşa Faaliyetleri ve Hamleleri", *Deniz Kuvvetleri Dergisi*, 74:460, 1968, pp. 3–11.

Özsaraç, Halil, *Donanmanın Tarihsel Serüveni*, Doruk Yayınları, İstanbul, 2020.

Piri Reis, *Kitabı Bahriye*, Türk Tarih Kurumu Yayınları, Ankara, 2002.

Ramis, O. & Y. Teofanidis, *Türk ve Yunan Deniz Harbi Hatıratı ve 1909–1913 Yunan Bahri Tarihi*, Büyük Erkânıharbiye XII. Deniz Şubesi, İstanbul, 1930.

Saffet, *Mezemorta Hüseyin Paşa*, (trans.) Yavuz Senemoğlu, Deniz Basımevi, İstanbul, 1994.

"Tarihçe", Türk Silahlı Kuvvetlerini Güçlendirme Vakfı, https://www.tskgv.org.tr/tr/hakkimizda/tarihce.

"TCG Heybeliada-(MİLGEM-Milli Gemi-National Ship) Patrol and Anti-Submarine Warfare Ship" https://www.globalsecurity.org/military/world/europe/tcg-milgem.htm.

Terzioğlu, Sait Arif, "Şilep Yapmak Muhrip Yapmaktan Kolaydır", *Derya Aylık Denizcilik Dergisi*, 56, 1972, pp. 10–11.

Tezel, Hayati, *Anadolu Türklerinin Deniz Tarihi*, Deniz Basımevi, İstanbul, 1973.

Toluner, Sevin, *Kıbrıs Uyuşmazlığı ve Milletlerarası Hukuk*, İstanbul Üniversitesi Yayınları, İstanbul, 1977.

Uzunçarşılı, İsmail Hakkı, *Osmanlı Devleti'nin Merkez ve Bahriye Teşkilatı*, Türk Tarih Kurumu Yayınları, Ankara, 1984.

Wicaksono, Tresno & Anak Agung Banyu Perwita, "The Military Industrial Complex in a Developing Country: Lessons from the Republic of Turkey", *Jurnal Hubungan Internasional*, 9:1, 2020, pp. 53–68.

Woods, Henry F., *Türkiye Anıları*, (trans.) Fahri Çoker, Milliyet Yayınları, İstanbul, 1976.

Yavuz, Celalettin, "Cemal Paşa'nın Almanya Gezisi: Bir Davetin Perde Arkası", *Atatürk Yolu*, 5:19, 1997, pp. 335–346.

Yavuz, Celalettin, "Sultan Abdülaziz Donanması: Yelkenli Teknelerden Buhar Makineli Gemilere Geçiş, Bitmeyen Reform İhtiyaçları", *XIII. Türk Tarih Kongresi*, Türk Tarih Kurumu, Ankara, 2002, pp. 1805–1837.

5 From Systemic Emulation to Military Innovation
Turkish Drones and International Politics

Baybars Öğün

Introduction

Türkiye's counter-terrorism operations both within and beyond its borders, as well as its active role in the Syrian and Libyan civil wars and its indirect effects in the Nagorno-Karabakh War have all involved the use of uncrewed aerial vehicles (UAVs) and uncrewed combat aerial vehicles (UCAVs), and this has garnered considerable international attention.[1] In addition, this issue has started to come up again with the videos of the Turkish military drones the Ukrainian Army has been using in the Russian invasion of Ukraine that are frequently published on social media.[2] American political scientist Francis Fukuyama focused on Türkiye's military drones and argued that Türkiye has recently surpassed the United States of America (USA), Russia, and China in terms of regional effectiveness.[3] Why has Türkiye's visibility in the international media increased dramatically in recent years? Are drone technology and its military use a new phenomenon in international politics? Was Türkiye the first to both produce and use these kinds of weapons in the world? What makes Türkiye different and unique in military-political terms? According to the distribution of capabilities in the international political system, can secondary states also create certain systemic consequences similar to great powers? Can Türkiye's production and use of military drones have various effects on international politics? How can this issue be addressed theoretically only at the system level? This chapter uses the conceptual and theoretical framework of Structural Realism, also known as Neorealism or Waltzian Realism, to seek answers to these questions.

Mainstream international relations schools explicitly take great powers as objects of study.[4] The main reason for this approach is that the source of many developments and interactions at the global level is related to these types of states. However, the pool created by secondary states (i.e., lesser or second tier states) is also quite large and requires analysis.[5] This topic has a very essential question. Is focusing on secondary states an inconsistency for Structural Realism? Although secondary states are not prioritized in Structural Realism, this chapter aims to emphasize the potential effects secondary states have within the framework of basic Waltzian concepts. Meanwhile, Türkiye constitutes a case in this regard, and Structural Realism

DOI: 10.4324/9781003327127-6

is not a Realist school that purports to explain foreign policy.[6] This chapter emphasizes system-level generalizations and repetitive behaviors according to the conceptual and theoretical framework of the Structural Realist school. It focuses not on Turkish foreign policy decisions but on some of the system-level effects Türkiye's various military practices have had.

The main argument of this chapter is that secondary states, like great powers, are able to develop imitable military practices, methods, or approaches. The emulation process has a nature that concerns all states. Imitable practices are associated with successful state practices. This situation can be handled in two ways. The first involves the incentive effect created by the states that perform military inventions. One example is the imitation of the atomic bomb and aircraft carriers, which are US invention.[7] The second involves the development of imitable military technology using new methods or practices. As an example, due to the spread of missile defense systems, imitator states have also produced this technology and adapted it to their structures. Israel's Iron Dome missile defense system[8] can be considered an excellent example in this regard. According to this approach, Türkiye's production and use of military drones fall into the second category. The inventor countries of UCAVs are accepted as Britain and the USA. Türkiye has imitated these weapons and used them with a unique military method. The effective use of Turkish-produced drones in preventive operations and heated conflicts at a level able to change the course of battles can create a demonstration effect for both great powers and other secondary states.

To reveal the theoretical framework of the chapter, the international political system requires definition. The chapter uses ideas from some scholars in this school, in particular those of Kenneth Waltz as the founder of Structural Realism. Accordingly, this chapter will discuss: (1) the definition of system; (2) the framework of structure-actor relations; (3) the differences between systemic effects and intra-system effects; (4) the structural motivations and tendencies of the states that make up the system; and (5) the main structural elements that direct states with different power distributions. The main interest of the chapter is to be able to explain how the system drives states and what their general tendencies are regarding structural constraints. From this point of view, one should remember that secondary states are also parts of the system. This chapter will also discuss the concept of emulation while accepting the general assumptions of Waltzian-based Structural Realism. In addition, this chapter will analyze: (1) the concept of innovation; (2) the relationship between emulation and innovation; (3) the concept of reinvention; (4) the differences between invention and reinvention; (5) the concept of adoption; and (6) the possibilities secondary states have to create a demonstration effect.

The Structure of the International Political System

Many system studies occur in the discipline of International Relations. However, their approaches to the system structure and the structure-unit relationship differ from one another.[9] The international system is usually analyzed in accordance with state-to-state interactions, crises, technical advancements, and global issues. As the levels of analysis change, so do the cause-effect relationships.[10] After Kenneth Waltz, the structure of the international political system began to be evaluated using a structuralist approach apart from states and interactions between states.[11] Although this may seem insignificant at first glance, this situation reflects a serious theoretical break in the discipline, according to which the structure of the system limits and indirectly controls states despite its unobservability. This study intends the international system to mean the international political system. The system of states historically corresponds to the post-Westphalian anarchic system.

Waltz was influenced by Emile Durkheim's approach, which he described as the resistance and pressure of social structures. Durkheim referred to social structures as consisting of individuals but also as having become phenomena that transcend individuals.[12] The international political system consists of states that existed before the structure (i.e., social structures) but then transformed from these units into an independent phenomenon. In addition, this independent phenomenon suppresses, affects, and resists existing units. Lastly, structure as a social object is unobservable.

According to Waltz, the structure produces restrictive circumstances and situations, such as in economic markets. While the market directs the firms, it encourages certain behaviors and punishes others. The market does not create the results as a direct decision-maker but does indirectly lead companies.[13] Accordingly, the structure of the system is independent of the parts or units that make up the system, their qualities, and their interrelations.[14] Robert Jervis stated that the system cannot be understood by summing up the relations or behaviors of the units within the system. Accordingly, the outputs of the system emerge independently from the interactions of the units.[15] In his early work, Stanley Hoffmann defined the system as an environment that both constrains and enables units.[16]

While domestic political systems are centralized and hierarchical, the international political system is decentralized. The units that make up the system (i.e., the states) are positioned horizontally. This means that no subordinate-superior relationship or chain of command exists among units. The international political system, having no supreme authority, also has no general framework regulating its units' duties or responsibilities. Each state is responsible for itself. In other words, the organizing principle of the international political system is anarchy.[17] The first consequence of the structure having an anarchic order is that *the state of nature is a state of war* between states. A state of war does not mean that wars and conflicts are constantly occurring, but that states are always ready for war in a horizontal position where no higher authority exists.[18] According to Tang, one of the main differences between anarchy and hierarchy is that anarchy creates a more ambiguous situation about the intentions and fears in the system. States in an anarchic system must always be prepared for the worst-case scenario.[19]

Waltz revealed the anarchic international political system to indirectly guide states in two ways: through socialization and through competition. According to this, states can survive to the extent that they are successful in adapting to the structural conditions of the system, and this is called socialization. Because the anarchic nature of the system leads states to a relative power struggle, another structural element involves the ability to emulate and imitate. Emulation, which is handled militarily in particular, also has political and economic reflections. Thus, while states try to adapt to the international political system in order to survive and increase their global or regional gains, they can also maintain their existence in the system by imitating the necessary military, political, or economic mechanisms that are in place at the time.[20]

Socialization

How does the structure socialize when it indirectly influences and constrains states' behaviors? In other words, how does the structure connect with the units? Waltz's answer is as follows:

> Structure affects behavior within the system but does so indirectly. The effects are produced in two ways: through socialization of the actors and through competition among them. These two pervasive processes occur in international politics as they do in societies of all sorts. Because they are fundamental processes....[21]

In the process of socialization, individuals tend to behave in accordance with the social structure. This situation occurs subconsciously and spontaneously. Individuals become compatible with the norms and rules of society. Thus, the integrity of the structure is preserved, and similar behaviors are produced because of socialization.[22] To understand socialization further, one may consider Waltz's best-known historical example. When the Bolsheviks pulled Russia out of World War I through the Brest-Litovsk Treaty dated March 3, 1918, the secret agreements Tsarist Russia had made were announced to world public opinion. This was the Bolsheviks' first explicit response to the rejection of established diplomatic methods. Despite declaring that their foreign policy would represent their principles after the October Revolution, Soviet statesmen quickly adopted diplomatic processes, behavior patterns, and political maneuvering within the international system. For example, when the Minister of Foreign Affairs, Chicherin attended the Genoa Conference in 1922, he was instructed to avoid making any pretentious remarks reflecting Bolshevik ideology.[23] This notice aimed to eliminate discourses that might contradict the attitudes that had become customary in diplomatic relations between states. States want to be rewarded by acting in accordance with the rules and practices within the group.[24] Thies also identified the essential elements of socialization as the rules, norms, principles, and conventions that form in the system.[25]

Through socialization, the anarchic structure of the international political system encourages units to be compatible with the structure. These socialization practices appear in the form of coercion or encouragement. Accordingly, states act according to the invisible constraints of the structure. States that are incompatible with the structure are excluded or penalized. When the common rules, reflexes, or actions arising from the coexistence of states are not complied with, the structure either forces the incompatible ones to become compatible or excludes them from the system. This state of exclusion can be understood most clearly through failed states. Socialization takes place in practice through the intervention of other states. Thus, invisible structural effects become visible through states' behaviors. For example, states must be legally and politically recognized by other states.[26]

Competition

According to Waltz, the second way the structure of the system relates to units is through competition. Competing states tend to imitate successful examples. Like socialization, this process takes place spontaneously and unconsciously. Behavioral imitation has firms in an economic market or states in an international political system become similar at the behavioral level.[27] The structural effects of both socialization of and competition among units is seen as a type of natural selection.[28] According to this, states that adapt to the structural orientations will continue to exist, while those that cannot keep up will be unable to avoid extinction. This also concerns the question of whether states are successful or not.[29] While socialization ensures that the units become compatible with one another, competition causes the weak units in the system to be eliminated. The effects of socialization and competition are observed in the form of emulation, imitation, and adoption of successful behaviors and the disappearance of unsuccessful units from the system. These effects emerge as states harmonize their domestic political, legal, and military mechanisms with the system.

States have to compete with others by focusing on their relative gains to survive. The existence of competition does not mean that cooperation mechanisms do not exist. States that have to fend for themselves are compelled to compete to survive. Cooperation mechanisms are part of this competition. The anarchic structure of the international political system punishes altruism. Cooperation between states is not about doing good for free but about states' struggling to exist alongside each other. Thus, the element of structural competition reveals a mechanism that operates in a loop. All states try to imitate technological developments, military innovations, foreign policy behaviors, and legal political practices. The most successful states are the most talented at imitation. The main feature of unsuccessful states in the system is their inability to compete due to their incapability of imitating. The basic concepts associated with the phenomenon of competition are emulation and imitation in states' external behaviors.[30]

Emulation

The emulation and innovation studies in the literature are generally observed to be actor-centered, with organization theory and cultural studies being particularly prominent.[31] The main reason for this is the belief that social dynamics and decision-making and bureaucratic processes need to be considered as the causal factors. As a result, the assumption of the rational and conscious actor comes to the fore. In addition, some approaches address emulation through states' learning processes, as is the case with socialization.[32] Structural Realists attempt to explain emulation and innovation at the system level. In these studies, the main factor is observed to be not the actor but the structure.

The basic feature in states' power struggles based on relative gains at the system level is that it reveals structural similarity. Thus, states attempt to adopt other states' successful practices and methods into their own structures to avoid falling behind. As such, emulation results in imitation. In other words, the essence of competition is imitation. According to Goldman and Andres, "Emulation refers to the imitation of the innovation".[33] Waltz expressed this situation as follows:

> Competition produces a tendency toward the sameness of the competitors. Thus, Bismarck's startling victories over Austria in 1866 and over France in 1870 quickly led the major continental powers (and Japan) to imitate the Prussian military staff system, and the failure of Britain and the United States to follow the pattern simply indicated that they were outside the immediate arena of competition. Contending states imitate the military innovations contrived by the country of greatest capability and ingenuity.[34]

Joao Resende-Santos defined emulation as follows:

> In the military sphere, emulation is the imitation of any aspect of another state's military system that bears upon the organization, strategy, or armaments of the imitator's own system, and which bring it into reasonably close similarity or correspondence with that of the state being emulated. The most frequent and widespread type of military emulation is the adoption of specific military technology.[35]

Barry Posen described the emulation created by competition based on relative gains as follows:

> Military capabilities are a key means to such security, and thus states will pay close attention to them. States will be concerned about the size and effectiveness of their military organizations relative to their neighbors. As in any competitive system, successful practices will be imitated.[36]

Competing states try not to fall behind in the relative power competition by imitating each other's successful military, political and economic practices.

For example, the Ottoman Empire initiated military reforms because of the great wars that had been lost, one after another, since the 18th century.[37] As a result of these reforms, innovations and practices in the European armies created radical changes in the Ottoman Army.[38]

Many European officers who fled their countries were integrated into the Ottoman army. After the 1848 Revolution in Europe in particular, Hungarian and Polish refugee officers took shelter in the Ottoman Empire and served in the Ottoman Army as Muslims.[39] This policy aimed to imitate successful European army practices and prevent weaknesses on the battlefield. After the Russo-Turkish War (1877–1878) ended in defeat, the most successful examples began to be sought for military reform.[40] A German influence can be seen in this regard. According to İlber Ortaylı, the general understanding of military strategy and discipline created a German sympathy during the Ottoman Empire that could be easily observed both in the Ottoman military-civil bureaucracy and in Sultan Abdülhamid II through the effect of Prussia's victory against France. As a result, German officers were contracted into the ranks of the Ottoman Army. They served as the commanders and trainers of the Turkish officers who would mark the last period of the Ottoman Empire.[41] Moreover, imitation of Prussian military techniques and practices in the 19th century was a phenomenon frequently observed from the Americas to the continent of Asia.[42] The tradition of military reform in the Ottoman Empire also pioneered political, legal, social, and economic reform initiatives in the later periods. All the steps taken for the survival of the state emerged in order to incorporate the successful practices of other states.

Resende-Santos described the modernization processes of the Ottoman Empire, Japan, Russia, Egypt, and France in the 19th century as "large-scale military emulation" because of their gradual and systematic nature.[43] However, Resende-Santos' research not only viewed the emulation process as a full military, political, or social modernization but also included short- and medium-term reflections of military advances. This is called "small-scale or one-time emulation". The vast majority of emulation processes which are definable as "the imitation of specific weapons technologies, equipment, uniforms, or discrete battlefield tactics" are discussed in this context.[44] Therefore, the military reforms of the Ottoman Empire and Türkiye's specific military weapons (i.e., UAVs and UCAVs) that are the subject of this chapter, are not evaluable at the same conceptual level.

Mainstream studies considered great powers' creation of an incentive effect at the system level to be usual. The main question here is whether secondary states might have a similar effect. Another question is what kind of emulation process secondary states might lead to. First of all, the literature implicitly states that "few of the most capable" great powers have this effect.[45] In other words, successful practices that lead states to imitation are naturally developed by the most capable powers. However, no theoretical obstacle is found for secondary states to create an incentive effect in the system by carrying out military innovations. These types of situations just occur less frequently. To explain this situation, emulation should be handled at the system level. Therefore, the analysis will be made over forms of balance that emerge according to the power distributions at the system level. This is because the different forms of balance (i.e., internal and external efforts)[46] the great powers apply over different power distributions also affect secondary states.

Emulation and the Distribution of Power

Waltz described the change in the structure of the international political system in accordance with the distribution of the most capable units in the system.[47] The prominent concept here is polarity, which refers to the power distribution in the system.[48] Therefore, polarity is a concept associated with the great powers. The main issue that Waltz focused on is the forms of balance that emerge between the great powers regarding different power distributions. Accordingly, the great powers' main form of balance in the multipolar system is external balancing, while in the bipolar system it is internal balancing.[49] Resende-Santos refers to emulation in place of internal balancing, while considering external balancing only as balancing.[50]

The structure of the system is defined by the number of poles. This figure determines the basic balancing act between great powers. Changes in the system-level balancing pattern affect states' behaviors. For example, the balance of power with regard to multipolarity is based on the external balancing between great powers, (i.e., alliances). Although forming alliances has become flexible, the increasing interdependence among states limits their ease to act. With regard to the bipolar system, on the other hand, the superpowers are not dependent on alliances with other states, as they apply internal balancing simply by relying on their own abilities and increasing their capacities. Thus, the level of interdependence between superpowers in a bipolar system is very low.[51]

How does polarity relate to the emulation process? According to Resende-Santos, imitation is a kind of internal balancing practice. Accordingly, the development of new weapon technologies and military methods by imitating each other in the relative power struggle should be evaluated within the scope of internal balancing.[52] For the diffusion of military innovations, systemic conditions must be created in which states will tend toward internal balancing. According to Goldman and Andres, external balancing slows down the emulation process.[53] In Structural Realism, the internal and external balancing among great powers is handled in accordance with the different power distributions. Therefore, the global balance of power affects how secondary states apply their forms of balance.

With regard to multipolarity, the uncertainty in the system is very high. Therefore, the emulation process becomes quite common. With regard to bipolarity, on the other hand, the emulation process occurs less frequently for other states, as the superpowers rigidly build the security structures of their own bloc.[54] Resende-Santos refers to Benjamin Frankel on this point, who focused on nuclear proliferation in different power distributions. Frankel compared the Cold War period and the post-Cold War period in terms of the proliferation of nuclear weapons and stated the following on the subject:

> The more important incentive for intensified proliferation, however, is the change in the international political system. The change that will most directly influence the acceleration of nuclear proliferation is the move away from bipolarity. The end of bipolarity means that superpower guarantees – the most effective instrument to moderate the effects of systemic characteristics – will be reduced and weakened. As a result, the international system will revert to a more unvarnished form of anarchy in which systemic attributes such as the security dilemma and self-help will be accentuated.[55]

When bipolarity is defined as two rival blocs, states in the same bloc are assumed to have similar foreign policy objectives.[56] This similarity is a reflection of the efforts great powers make to surround each other at the global level. As a result, the emulation processes of secondary states slow down in the bipolar system. According to Waltz, the international political system, being a self-help system, can be defined when there are at least two great powers. Thus, according to this logic, the bipolar system corresponds to the most stable system. As the number increases, the function of competition and bargaining changes, relationships become more complex, and costs increase. As the number decreases, the functioning of the system becomes predictable.[57]

The weight internal balancing has in the bipolar system is more understandable through the example of NATO and the Warsaw Pact. Both the USA and the USSR wanted to increase the number of their allies and share their burdens with them. However, this expectation of contribution is not as important as it is in the multipolar system. The USA and the USSR were not dependent on their allies because of their power capabilities during the Cold War. For example, the withdrawal of France from the military wing of NATO during the Charles De Gaulle period did not cause any change in the balance of power.[58] In the bipolar system, the dependence of other states on the superpowers emerges, rather than interdependence. As such, the process of internal balancing or emulation is handled between superpowers. From this point of view, the possibilities for applying secondary states' internal balancing with regard to different power distributions turn out to be different. In the bipolar system, secondary states' emulation processes are slower as these states guarantee their security under the superpowers' blocs. The main reason for this is the stability of the system.

One of the differences between multipolar and bipolar systems is understood to be the level of uncertainty within the system. The superpowers in the bipolar system are very sensitive to threats. The source of threats for the USA was the USSR, and the source of threats for the USSR was the USA. Thus, no matter where in the world, any development, issue, or crisis concerned these two superpowers. In this system, one party's gain is seen as the other's absolute loss. Therefore, certainty is present regarding how the balance of power in the system works. Miscalculations occur based on external balancing in the multipolar system, while in the bipolar system, sudden and disproportionate reactions based on internal balancing create danger.[59]

As a result, secondary states' follow-up of emulation processes slows down at a systemic level in the bipolar system, while these states are able to act more autonomously in the multipolar or unipolar system. Waltz stated that the global balance of power is carried out by at least two states and argued a transition period would occur based on the US alone after the Cold War. Unipolarity would eventually evolve into a bipolar or multipolar system in accordance with these assumptions.[60] Therefore, the structural constraints of the bipolar system can be said to have disappeared because a transition period occurred in the post-Cold War system. Accordingly, secondary states could imitate successful military technologies faster, as envisaged in the multipolar system.

The Scope of Military Innovation

Emulation is related to imitation, adoption, innovation, invention, reinvention, and the demonstration effect. When the relationship between these concepts is established consistently, the status of the secondary states in the system can be explained more easily. The literature generally considers the concept of innovation as a change in military technology, new military methods, approaches, or organizational forms in specific fields. In addition, military innovation is usually limited to the applications from World Wars I and II and the Cold War period.[61] Therefore, the literature focuses on the scope of and reasons for military innovation in line with the needs of great powers. The nature of innovation, the classification of states according to their position in the system, and the systemic consequences involving secondary states are ignored.

"A new approach to operations"[62] is the expression Farrell used while revealing the general scope of innovation, and it identifies the current study's field of interest. In this way, innovation is handled outside of a system-level global war or the practices of great powers. Essentially, the practices of Türkiye that have attracted attention in the international community are how it has produced its own military technology as well as combined it, using the new approach Türkiye has brought to military operations.[63] To clearly demonstrate the concept of innovation, certain words need to be distinguished that have been used alongside innovation. Some ambiguity exists in the literature regarding the terms "innovation", "invention", and "reinvention" as well as the distinction between "innovation" and "adoption". Rogers summarized this as follows:

> Most scholars in the past have made a distinction between invention and innovation. Invention is the process by which a new idea is discovered or created, while adoption is a decision to make full use of an innovation as

the best course of action available. Thus, adoption is the process of adopting an existing idea. This difference between invention and adoption, however, is not so clear-cut when we acknowledge that an innovation is not necessarily a fixed entity as it diffuses within a social system. For this reason, "reinvention" seems like a rather appropriate word to describe the degree to which an innovation is changed or modified by the user in the process of its adoption and implementation.[64]

When examining these concepts, an invention must clearly occur first. Consumers of a product are formed when a new idea is put into practice and turned into that tangible product. This process can be called adoption when the product is fully imitated and used. In other words, adopting an imitated new product or method is a process. Imitation temporarily occurs first, followed by adoption. Meanwhile, reinvention involves how users or consumers develop this product in line with their understanding when adopting the imitated technology or product. Although the nature of innovation does not change as a result of reinvention, an innovation is revealed in its application or method.[65] This is the difference between the concept of reinvention and invention. Thus, the unique skill of the consumer who is not the innovator of the product emerges in the context of reinvention. In the framework of military innovation, reinvention is considered for developing a brand-new weapon and adapting new technology to its structure.[66]

When considering innovation as the upper concept, invention and reinvention become its sub-concepts. In this way, these two processes are seen to be the main elements of the innovation process when making a distinction between invention and reinvention. Accordingly, the relationship between the concepts is revealed in Figure 5.1:

In competition, the structure of the system directs all states to the emulation process in order to survive. The results from emulation are to imitate and

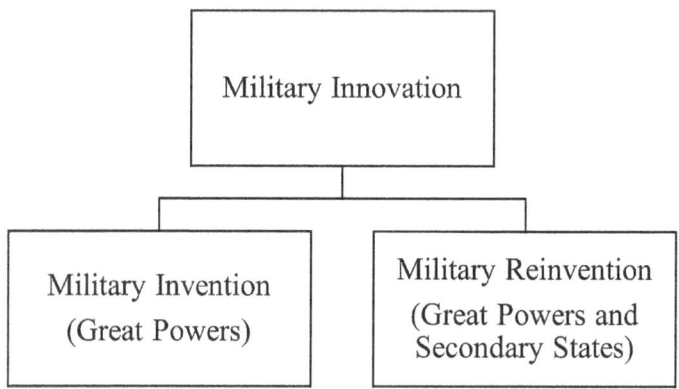

Figure 5.1 The Concept of Military Innovation.

Sources: *Created by the author.*

to adopt. This process is passive. On the other hand, innovation is not a process that every state can manage. Only states with certain capabilities can innovate, as innovation is an active process. Successful state practices are imitated because of emulation. However, their successful adoption is not guaranteed. Not every state can compete perfectly. Although the system may have such a structural orientation, the results may not always be successful. However, these different results cannot be explained at the system level. Innovation may arise as brand-new products and methods in the form of military inventions from the great powers. However, evaluating reinvention in terms of secondary states' ability to develop their new methods in the emulation process is also appropriate. This situation represents the stage where emulation turns into innovation. Emulation and innovation processes can be compared as follows (Figure 5.2):

The fact that secondary states create an imitation effect in the system is related to the concept of reinvention. Of course, secondary states may also

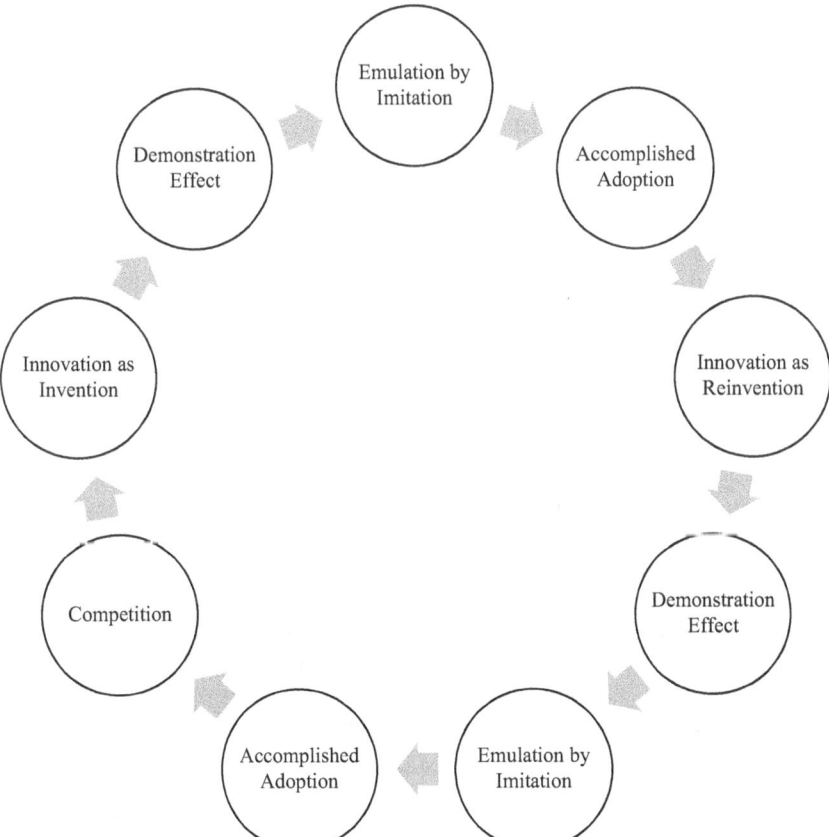

Figure 5.2 The Cycle of Emulation to Military Innovation.

Sources: *Created by the author.*

have inventions. However, looking at their position in the system, inventions are seen to appear in proportion to the capabilities of the great powers. On the other hand, secondary states can create demonstration effects in the system. Handling of an imitated military technology in the emulation process with a new method, especially during the implementation phase, may attract the attention of other states. Therefore, the concept of the demonstration effect also emerges at this phase. The demonstration effect is when the success of a state in the reinvention process creates an incentive effect on other states. Buzan, Jones, and Little talked about the demonstration effect from inventing methods and technologies as ones that states can emulate:

> Socialization works by the demonstration effect of the most successful units, which encourages others to copy them. In the postwar period, the former Soviet Union has been a model for some, the United States for others, and more recently, Japan has set the pace as a model of economic efficiency and prowess. The collapse of communism in 1989 not only removed one demonstration model from the system, but also illustrated the powerful socialization effect of more successful units as much of the ex-communist world struggled to adopt the forms and practices of the West. The effect of competition is rivetingly demonstrated by the period of European imperialism.[67]

Türkiye has arrived at a position where it can produce a technology that other states have previously developed. Therefore, Türkiye's production of military drones itself is not an invention. Domestic production is essential in terms of reducing Türkiye's dependence on foreign weapons and facilitating its ability to take the initiative in foreign policy. However, the main difference Türkiye has created is its field application of this technology with a brand-new approach, by offering a novel method to use these weapons most effectively in terrorist operations and conventional warfare.[68] Therefore, after a certain period it would be no surprise if other states attempt to use UAV technology in a similar way to Türkiye. This would be a natural consequence of systemic emulation that fits into the concept of reinvention. Türkiye's performance in military operations through military drones emerges as a method that other states will also want to implement. According to Francis Fukuyama, the military outcomes of Türkiye's unique employment of drone technology may have a transformative influence in ground warfare.[69] The demonstration effect allows successful states to be distinguished. A state that produces military technology or develops a new method increases its autonomy and directs other states to imitate it. The most harmonious and successful states in the system are distinguished in this way.

The Structural Constraints on Türkiye During the Cold War

Türkiye's foreign policy and military capability in the bipolar system can be analyzed within the scope of a unilateral dependency relationship within the

NATO alliance. The main reason for this is that Türkiye was heavily affected by the restrictions of the system, which compels secondary states to be dependent on the blocs. As a result, not only did Türkiye buy American-made weapons, but it also had to make commitments about how, when, and where these weapons would be used. The most prominent examples were the Jupiter missile crisis, the U2 incident, the Johnson Letter, and the 1975 American arms embargo.

The first example is about the relationship between the Jupiter missiles and the Cuban Missile Crisis. With the USSR's launch of rockets into space (Sputnik I) and a satellite into orbit (Sputnik II) in 1957, the USA's internal balancing practices combined with containment strategies. The Eisenhower Doctrine, proclaimed in 1952, pertains to an all-out struggle using nuclear weapons against a possible Soviet attack. For this reason, the doctrine planned to place nuclear warhead missiles in certain NATO member countries close to the USSR. With the approval of Türkiye and Italy, medium-range and nuclear-armed Jupiter missiles were placed in these two countries in accordance with this. Through the secret agreement made in 1959, 15 Jupiter missiles were deployed to Türkiye and made ready for use in 1962.[70] However, this had unwittingly made Ankara one of the hidden actors of a global crisis. After the USA deployed medium-range Jupiter missiles to Türkiye, the USSR imitated this action as the product of its nuclear containment strategy and placed its own medium-range and nuclear warhead missiles in Cuba in 1962. The USA besieged Soviet ships carrying these missiles' launch pads on October 22 before reaching Cuba, and thus the famous Cuban Missile Crisis emerged. The Moscow administration wanted the missiles to be withdrawn from Türkiye as a condition for removing the missiles from Cuba. Despite Türkiye's opposition to the withdrawal of the missiles, the decision was made to dismantle the Jupiter missiles after secret negotiations between the USA and the USSR. Türkiye was informed, as the reason for the withdrawal, that the missiles were obsolete and that Polaris nuclear submarines would be allocated instead. Türkiye has claimed to have been unaware of these secret bargains.[71]

The second example concerns the problems created by the use of American military bases in Türkiye. The USA's use of Turkish bases and the weapons deployed in military facilities caused a dilemma because ensuring Türkiye's security against the USSR had actually increased the security threat from the USSR. One of the most well-known examples of this was the U-2 issue. On May 1, 1960, an American U-2 spy plane, that had taken off from Incirlik base, was shot down by the USSR while it was collecting intelligence by entering Soviet airspace over Pakistan. Because the pilot had been captured alive, the USA was unable to cover up the issue.[72] Again in 1965, another U-2 plane that had taken off from Incirlik was shot down by the USSR, this time over the Black Sea. These flights were completely abolished as a result of pressure on Türkiye from the USSR.

The third example relates to one of the most well-known examples of the restriction on Türkiye's foreign policy behavior. The conflicts initiated by the

Greek Cypriots in August 1964 led to Türkiye's intervention as a guarantor country. In this environment, US President Johnson sent a letter to Turkish Prime Minister İsmet İnönü on June 5 making very harsh statements. The letter briefly stated that Türkiye could not use the military technology and weapons that it had received under the current agreement because of the Cyprus issue. In addition, the most important point for Türkiye's security was the emphasis on the uncertainty of NATO's defense of Türkiye in the face of a possible attack by the USSR.[73]

The last example is about the problems Türkiye has experienced with regard to arms purchases; an issue which is still ongoing. On July 15, 1974, a coup was staged against President Makarios in the Republic of Cyprus, that brought into power Nikos Sampson, who was known for his hostility to Turks. Türkiye perceived this initiative as an intervention within the framework of Enosis and proposed a joint intervention with the UK; however, this proposal was rejected. The Cyprus Peace Operation was launched on July 20, 1974 (First Operation). A ceasefire was declared on July 22, and the parties agreed in Geneva that the 1960 Treaty of Guarantee should remain in effect. The second Geneva negotiations started on August 8 and proceeded with no result. Türkiye then carried out the second operation on August 14. On August 16, the intervention ended, and the Cyprus issue evolved into its current state. The USA, while not reacting to the first intervention, suspended American aid to Türkiye on February 5, 1975, as a result of the second.[74]

The Turkish Reinvention of UAVs and UCAVs

With the end of the Cold War, the intense bloc pressure on the secondary states in the system disappeared. The alliance dependency, which was the product of the American containment strategy in the bipolar system, had loosened. As a result, more capable secondary states became able to apply internal balancing, especially those struggling for regional hegemony. Many secondary states worldwide are prone to internal balancing to the extent that bipolarity is not observed. Although Türkiye is still a NATO member, it has been able to increase its capabilities with its own means in order to create space and maneuverability for its foreign policies'. The changing power distribution in the system was observed to have encouraged Türkiye in this direction. Türkiye's technology and weapon production in recent years is evaluated within this framework. In this context, domestic UAVs and UCAVs have been developed rapidly, making it easier to achieve military-political results. Moreover, Türkiye has changed the course of the conflicts in which it has been engaged through UCAVs within the scope of reinvention. This is the most important difference Türkiye has made in the field. In addition to producing its UAVs and UCAVs domestically, Türkiye also offers a new method both for fighting terrorism and increasing its regional autonomy. Furthermore, Türkiye has started to export its UCAVs to Qatar and Ukraine,[75] with its weapons supply to NATO member Poland being particularly significant.[76]

Türkiye has been attempting to adapt drone technology to its structure since the late 1980s. The Turkish Armed Forces first started using the BTT-3 Banshee system produced by the company Meggitt in 1989 as its first UAV. In 1993, five Canadair-produced CL-89 UAVs were donated by Germany and started being flown in 1994. Afterward, US- and Israeli-made UAVs were chosen. Türkiye purchased the American-made GNAT and GNAT-I UAVs and Israeli-made Herons. Although Türkiye used these weapons for a while, various problems were experienced. In addition, the fluctuating course of Türkiye-Israel relations has negatively affected the arms trade. The main reason why Türkiye wanted to buy these weapons was to eliminate intelligence deficits, especially in military operations against the PKK, a designated terrorist organization.[77]

Having been dependent on arms imports in the past, Türkiye's most fundamental desire for change today is to produce its own military technology to a certain extent. In addition, Türkiye is trying to develop its own operational methods within the scope of military innovation. The concept of reinvention becomes necessary at this point. What does Türkiye do differently within the scope of reinvention? What are the differences between Türkiye's military practices and those of other states? First, Türkiye has added something new to the anti-terrorism capability it has been developing and continuously making progress with since the 1990s. Among the states that have struggled with asymmetric threats throughout world history, Türkiye is one of the few that has been able to achieve successful results with its military operations. Türkiye's successful military practices greatly impacted the process that resulted in the capture of PKK leader Abdullah Ocalan in 1999. Türkiye was able to increase diplomatic pressure over Russia, Italy, Greece, and Syria in this way, and the conditions under which the USA handed Ocalan to Türkiye were met. As a result, the power of the terrorist organization on the field has been severely curtailed, and the PKK's activity levels up to the present have been considerably reduced. Therefore, the PKK has had to give priority to organizing outside Türkiye's borders, and has attempted to maintain its existence under different names in northern Iraq since the 2003 Iraq War and in northern Syria since the Syrian civil war that started in 2011.

Türkiye has started to carry out cross-border military interventions against the terrorist groups along its own borders and against the de facto positions of terrorist organizations in northern Syria along the region shared with the Turkish border. The first of these counter-terrorist operations was called Euphrates Shield and launched on August 24, 2016. ISIS and PKK/YPG terrorist elements were cleared during the seven-month intervention in the area stretching from Jarablus to Al-Bab.[78] Operation Afrin (also called Olive Branch) was launched on January 20, 2018, against ISIS and PKK/YPG terrorist organizations in the Afrin region of Syria.[79]

Türkiye's military drones have increased its ability to strike along Türkiye's borders and beyond in asymmetric warfare. While carrying out intelligence activities and military operations with mostly ground forces from the Turkish Armed Forces (TAF) was traditionally a costly option, the Turkish UCAVs have allowed the entire course of asymmetric warfare to be carried out at low cost (in terms of money, ammunition, and loss of life). Türkiye, which reserves its right to self-defense in the fight against terrorism, gathers intelligence with military drones, scans the land, and conducts reconnaissance and tracking. With these, Türkiye is able to organize attacks on high-value targets. In addition, Turkish-type UCAVs are used with a unique military tactic. Therefore, the adversarial targets, including heavy weapons (e.g., tanks, howitzers, armored carriers), can be neutralized using military drones.

These operations mark the first time that Türkiye has carried out such extensive cross-border operations since Cyprus and northern Iraq. However, the US attitude toward the PKK/YPG terrorist elements has restricted Türkiye. On the other hand, Türkiye has demonstrated its lack of intention to harm the territorial integrity of Syria by carrying out joint patrols with Russia. Although all these restrict Türkiye in certain respects, it retains an active stance toward preventing the security crises it experienced after the First Gulf War. Accordingly, the secondary states' expansion of regional initiatives is the product of structural orientations. Reduced structural constraints in the system allow secondary states to diversify their security policies and implement internal balancing.

Türkiye used UCAVs for the first time in Operation Olive Branch in Syria in 2018. However, this usage has remained very limited. According to some experts, the difference Türkiye made militarily started with Operation Spring Shield. TAF used UCAVs in this operation with an unprecedented method. Accordingly, many UCAVs were flown together in a very narrow area simultaneously, and an all-out attack was carried out on heavily armed units. Thus, hundreds of tanks, armored vehicles, howitzers, and multi-barrel rocket launchers belonging to the regime were destroyed in Idlib. In addition, carrying out the attack in coordination with Turkish UAVs eliminated the intelligence deficit.[80] Ridvan Bari Urcosta expressed the military method Türkiye offered for the first time within the borders of Syria as follows:

> The Idlib campaign (Operation Spring Shield) was the first time Turkey had used its UAVs at such a massive scale and against a foreign country with as powerful a backer as Russia. During Spring Shield, Turkish UAVs were operating almost everywhere in the greater Idlib area and reached the deep rear of the Syrian Army. The penetration into the Syrian rear had serious psychological and military consequences. The Syrians spotted Turkish UAVs in Hama and Aleppo, territories under Syrian government control. In Idlib, the Turkish Army employed new drones for the first time, field testing its ANKA-S and Bayraktar-TB2 with intensity. Aside from traditional strategic or tactical roles, the UAVs were used to conduct so-called "sniper" missions, liquidating targeted groups and specific persons of interest. For example, Turkish UAVs reportedly liquidated two Syrian brigadier generals, a colonel, and foreign fighters from Hezbollah and Iran in an attack on Syrian headquarters in Zerba, south of Aleppo. Furthermore, Ankara actively promoted

Turkey as the first country to employ sophisticated small drones as a swarm in combat. Turkish officials claimed that this military innovation demonstrated Ankara's technological prowess on the battlefield. These swarms of remotely-controlled drones destroyed Syrian bases and chemical warfare depots, as well as air-defence systems.[81]

Türkiye was indirectly influential in the Nagorno-Karabakh War through the UAVs and UCAVs it allocated to the Azerbaijani Army. Azerbaijan has been modernizing its military for over a decade. The new weapon systems and military training exported from Israel and Türkiye increased their effectiveness. For example, Israeli-made Harop drones were also used in this war. The Turkish-made UAVs and UCAVs the Azerbaijani Army used in the Karabakh War were claimed to have been directed by Turkish military officials.[82] Although Azerbaijani President Ilham Aliyev rejected the view that Türkiye had been directly involved in the war, he stated that the UCAVs received from Türkiye had changed the course of the war.[83] Military experts stated that the drones easily hit their targets in this war, unlike the warplanes that had difficulty maneuvering due to geographic obstacles. The Bayraktar TB-2 drones, with their long flight time, ensured infiltration behind the defense lines of the Armenian troops.[84]

Türkiye intervened in the Libyan civil war by supporting the Government of National Accord (GNA), which the UN recognizes Libya's legitimate government. Starting in 2020, opposition forces led by Khalifa Haftar began to lose ground against the GNA because of Türkiye's active engagement using its UCAVs; this changed the course of the civil war.[85] Türkiye's active intervention in the Libyan civil war is related to the competition over energy resources in the eastern Mediterranean. A maritime rights deal between Türkiye and the GNA was made a few months before Turkish drones participated in the military operations; it could potentially expand Türkiye's maritime claims in waters currently contested by Greece.[86] Türkiye demonstrates a military method of achieving effective results at minimal costs within the scope of reinvention. This usage can in the short and medium term transform into a practice that other states will imitate. Therefore, it constitutes an important example in terms of realizing the military innovation element is imitated within the scope of reinvention, not invention. Türkiye's breakthroughs within the scope of military reinvention have been observed to help increase its effectiveness with regard to foreign policy.

Conclusion

This chapter has reconsidered the relationship between emulation and military innovation based on the conceptual and theoretical framework of Structural Realism. Due to classical studies between system-level emulation and innovation being insufficient in explaining secondary states, this study has constructed an innovation cycle. After classifying innovation as invention and reinvention, this chapter revealed the system-level innovation cycle through competition and emulation as follows: The structure of the international political system directs all states to survive in a relative power competition. Accordingly, competition requires being stronger than other states as much as possible. The first systemic result of competition is military innovation and developing a military technology for the first time is called an invention and is within the scope of innovation. However, invention is a process that is compatible with the capabilities of the great powers. As a result of an invention, a demonstration effect occurs in the system. Thus, great powers and secondary states try to imitate and adopt this technology. Some states that have successfully adopted this technology may use it in their own unique way, with different methods or approaches. Thus, reinvention occurs within the scope of innovation. Consequently, the demonstration effect recurs in the system yet again. Other states try to adopt this new method or approach by imitating it. This is how the cycle of military innovation continues. The relative power competition in the system also gets re-updated this way.

Türkiye's production of military drone technology and its active use in military operations have attracted the attention of the international public in recent years. The use of military drones is considered a military innovation of Türkiye, and not an invention. This weapon system should be evaluated within the scope of reinvention, which can be regarded as the development of a method by using a previously produced military technology. Nevertheless, it may also be considered within the scope of military innovation. Türkiye's effective use of military drones has not only strengthened its hand in foreign policy and the fight against terrorism, but also reduced its dependence on weapon imports. In addition, Türkiye's military reinvention at the system level is claimed to be able to create a demonstration effect and lead other states to emulate. According to this study's theoretical approach, other states are expected to imitate Türkiye due to the military innovation cycle.

This chapter has additionally argued systemic change to be the main reason secondary states turn to military reinvention. Why has Türkiye only recently been able to successfully implement military innovation and not during the Cold War years? According to the theoretical framework of Structural Realism, the restrictive effects of the bipolar international political system made implementation of internal balancing very difficult for secondary states. Secondary states had great difficulty avoiding the bloc policies the superpowers established in the Cold War. The crises that Türkiye experienced during that period clearly verify this situation. However, structural constraints on secondary states decreased once the bipolar system ended. Thus, states became able to turn to the emulation process, with Türkiye's attempts to develop its own defense industry being an indicator of this situation.

Notes

1 Samuel Bronsword, "Turkey's unprecedented ascent to drone superpower status", https://dronewars.net/2020/06/15/turkeys-unprecedented-ascent-to-drone-superpower-status/, accessed on 27.06.2021. Dan Sabbagh and Bethan McKernan, "Revealed: how UK technology fuelled Turkey's rise to global drone power",

https://www.theguardian.com/news/2019/nov/27/revealed-uk-technology-turkey-rise-global-drone-power, accessed on 24.06.2021. Kareem Fahim, "Turkey's military campaign beyond its borders is powered by homemade armed drones", https://www.washingtonpost.com/world/middle_east/turkey-drones-libya-nagorno-karabakh/2020/11/29/d8c98b96-29de-11eb-9c21-3cc501d0981f_story.html, accessed on 24.06.2021. James Marson and Brett Forrest, "Armed Low-Cost Drones, Made by Turkey, Reshape Battlefields and Geopolitics", https://www.wsj.com/articles/armed-low-cost-drones-made-by-turkey-reshape-battlefields-and-geopolitics-11622727370, accessed on 25.06.2021. Shaul Shay, "The Important Role of Turkish Drones in the Libyan War", https://www.israeldefense.co.il/en/node/39539, accessed on 26.06.2021.

2 "How useful are Turkish-made drones fighting in Ukraine?", https://www.dw.com/en/how-useful-are-turkish-made-drones-fighting-in-ukraine/a-61035894, accessed on 14.03.2022. Dave Philipps and Eric Schmitt, "Over Ukraine, Lumbering Turkish-Made Drones Are an Ominous Sign for Russia", https://www.nytimes.com/2022/03/11/us/politics/ukraine-military-drones-russia.html, accessed on 14.03.2022. "Turkish drone success in Ukraine sets stage for Asia roadshow", https://www.ft.com/content/03812d1f-6fce-4191-a3c3-0f87afbabc83, accessed on 04.04.2022.

3 Francis Fukuyama, "Droning On in the Middle East", https://www.americanpurpose.com/blog/fukuyama/droning-on/, accessed on 10.05.2021.

4 For some studies on this topic see John J. Mearsheimer, *The Tragedy of Great Power Politics*, W. W. Norton & Company, New York, 2001. Paul Kennedy, *The Rise and Fall of the Great Powers: Economic Change and Military Conflict from 1500 to 2000*, Random House, New York, 1987. Barry Buzan, *The United States and the Great Powers: World Politics in the Twenty-First Century*, Polity, Cambridge, 2004. Robert I. Rotberg and Theodore K. Rabb (Eds.), *The Origin and Prevention of Major Wars*, Cambridge University Press, Cambridge, 1989. Jack S. Levy, *War in the Modern Great Power System, 1495–1975*, The University Press of Kentucky, 1983. Joseph S. Nye, Jr., "The Changing Nature of World Power", *Political Science Quarterly*, 105:2, 1990, pp. 177–192. Tudor A. Onea, *The Grand Strategies of Great Powers*, Routledge, London, 2020. Håkan Edström and Jacob Westberg, *Military Strategy of Great Powers: Managing Power Asymmetry and Structural Change in the 21st Century*, Routledge, London, 2021.

5 There are many different approaches to classifying states. The most preferred ones about secondary states in the literature are small or middle powers. See Laurent Goetschel (ed.), *Small States Inside and Outside the European Union: Interests and Policies*, Kluwer Academic Publishers, Dordrecht, 1998. Robert O. Keohane, "Lilliputians' Dilemmas: Small States in International Politics", *International Organization*, 23:2, 1969, Spring, pp. 291–310. Andrew F. Cooper, Richard A. Higgott, and Kim Richard Nossal, *Relocating Middle Powers: Australia and Canada in a Changing World Order*, UBC Press, Vancouver, 1993. Eduard Jordaan, "The Concept of a Middle Power in International Relations: Distinguishing Between Emerging and Traditional Middle Powers", *Politikon*, 30:1, 2003, pp. 165–181.

6 For a discussion of whether the Structural Realist school is suitable for foreign policy studies, see. Colin Elman, "Horses for Courses: Why not Neorealist Theories of Foreign Policy?", *Security Studies*, 6:1, 1996, pp. 7–53. Kenneth Waltz, "International Politics is not Foreign Policy", *Security Studies*, 6:1, 1996, pp. 54–57.

7 See Joseph Cirincione, *Bomb Scare: The History and Future of Nuclear Weapons*, Columbia University Press, New York, 2007. Paul Fontenoy, *Aircraft Carriers: An Illustrated History of Their Impact*, ABC-CLIO, Oxford, 2006.

8 Arnon Gutfeld, "From 'Star Wars' to 'Iron Dome': US support of Israel's Missile Defense systems", *Middle Eastern Studies*, 53:6, 2017, pp. 934–948.

9 For the main systems studies in the discipline of International Relations, see. Morton A. Kaplan, *System and Process in International Politics*, John Wiley & Sons, New York, 1957. Richard Rosecrance, *Action and Reaction in World Politics: International Systems in Perspective*, Little, Brown, Boston, 1963. Stanley Hoffmann, *The State of War: Essays on the Theory and Practice of International Politics*, Pall Mall Press, New York, 1965. Charles A. McClelland, *Theory and the International System*, Macmillan, New York, 1966. George Modelski, *Long Cycles in World Politics*, Macmillan, London, 1987.

10 Kenneth Waltz, *Man, The State and War: A Theoretical Analysis*, Columbia University Press, New York, 1959. J. David Singer, "The Level-of-Analysis Problem in International Relations", *World Politics* (The International System: Theoretical Essays), 14:1, 1961, pp. 77–92.

11 Kenneth Waltz, *Theory of International Politics*, Addison-Wesley, Massachusetts, 1979.

12 Emile Durkheim, *The Rules of Sociological Method*, (trs.) W.D. Halls, The Free Press, New York, 1982.

13 Waltz, *Theory of International Politics*, pp. 73–74.

14 Waltz, *Theory of International Politics*, p. 40.

15 Robert Jervis, *System Effects: Complexity in Political and Social Life*, Princeton University Press, 1997, p. 6.

16 Stanley Hoffmann, "An American Social Science: International Relations", *Daedalus*, 106:3, 1977, p. 53.

17 Waltz, *Theory of International Politics*, p. 88. K. J. Holsti, *International Politics: A Framework for Analysis* 7th Edition, Prentice-Hall International Editions, New Jersey, 1995, p. 5.

18 Waltz, *Theory of International Politics*, p. 102.

19 Shiping Tang, "Fear in International Politics: Two Positions", *International Studies Review*, 10:3, 2008, p. 467.

20 Waltz, *Theory of International Politics*, pp. 74–77.

21 Waltz, *Theory of International Politics*, p. 74.

22 Waltz, *Theory of International Politics*, pp. 75–76. See about homogenization issue: Alastair Iain Johnston, "Treating International Institutions as Social Environments", *International Studies Quarterly*, 45:4, 2001, pp. 489–490.

23 Waltz, *Theory of International Politics*, pp. 127–128.

24 Frank Schimmelfennig, "International Socialization in the New Europe: Rational Action in an Institutional Environment", *European Journal of International Relations*, 6:1, 2000, p. 117.

25 Cameron G. Thies, "State Socialization and Structural Realism", *Security Studies*, 19:4, 2010, p. 694. Some studies consider the socialization process of actors as a process of learning and transferring what has been learned, based on the rational actor assumption. See G. John Ikenberry and Charles A. Kupchan, "Socialization and Hegemonic Power", *International Organization*, 44:3, 1990, p. 289.

26 For some studies on the role of international organizations and norms in adapting states to the system, see: Brian Greenhill, "The Company You Keep: International Socialization and the Diffusion of Human Rights Norms", *International Studies Quarterly*, 54:1, 2010, pp. 127–145 and Judith Kelley, "International Actors on the Domestic Scene: Membership Conditionality and Socialization by International Institutions", *International Organization*, 58:3, 2004, pp. 425–457.

27 Waltz, *Theory of International Politics*, pp. 76–77.

28 Jervis, *System Effects: Complexity in Political and Social Life*, p. 104.

29 Nicholas Onuf, "Structure? What Structure?", *International Relations*, 23:2, 2009, p. 191.

30 Ewan Harrison, *The Post-Cold War International System: Strategies, Institutions and Reflexivity*, Routledge, New York, 2004, p. 30.

31 See Barbara Levitt and James G. March, "Organizational Learning," *Annual Review of Sociology*, 14, 1988, pp. 319–340. Paul J. DiMaggio and Walter W. Powell, "The Iron Cage Revisited: Institutional Isomorphism and Collective Rationality in Organizational Fields", *American Sociological Review*, 48, 1983, pp. 147–160. Theo Farrell, "Culture and Military Power", *Review of International Studies*, 24:3, 1998, pp. 407–441. Theo Farrell, "World Culture and Military Power", *Security Studies*, 14:3, 2005, pp. 448–488.

32 See Benjamin Goldsmith, "Imitation in International Relations: Analogies, Vicarious Learning, and Foreign Policy", *International Interactions*, 29:3, 2003, pp. 237–267.

33 Emily O. Goldman and Richard B. Andres, "Systemic Effects of Military Innovation and Diffusion", *Security Studies*, 8:4, 1999, p. 81.

34 Waltz, *Theory of International Politics*, p. 127.

35 Joao Resende-Santos, "Anarchy and the Emulation of Military Systems: Military Organization and Technology in South America, 1870–1930", *Security Studies*, 5:3, 1996, p. 199.

36 Barry R. Posen, "Nationalism, the Mass Army, and Military Power", *International Security*, 18:2, 1993, p. 82.

37 The Westernization of the Ottoman Empire is related to its imitation of Western practices in the areas where it fell behind. The process of Westernization began to take shape after the 1699 Karlowitz and 1718 Passarowitz Treaties, that is, after the great wars in which land was lost for the first time.

38 The reforms initiated by the Ottoman Empire in the army became the driving force of Ottoman modernization. Modernization has also spread to cultural and social life over time and has been shaped as a motto not to be left behind from Western civilization. See İlber Ortaylı, *İmparatorluğun En Uzun Yüzyılı*, 25th Ed., Alkım, Istanbul, 2006, p. 43.

39 Ortaylı, *İmparatorluğun En Uzun Yüzyılı*, p. 44.

40 Mesut Uyar and Edward J. Erickson, *A Military History of the Ottomans: From Osman to Atatürk*, ABC-CLIO, California, 2009, p. 197.

41 İlber Ortaylı, *Osmanlı İmparatorluğu'nda Alman Nüfuzu*, 11th Ed., Timaş, Istanbul, 2010, pp. 85–86. Additionally see Serhat Guvenc & Mesut Uyar, "Lost in translation or transformation? The impact of American aid on the Turkish military, 1947–60", *Cold War History*, 22:1, 2022, pp. 60–62.

42 John A. Lynn, "The Evolution of Army Style in the Modern West, 800–2000", *The International History Review*, 18:3, 1996, p. 510.

43 Resende-Santos, "Anarchy and the Emulation of Military Systems: Military Organization and Technology in South America, 1870–1930", pp. 200–201.

44 Joao Resende-Santos, *Neorealism, States, and the Modern Mass Army*, Cambridge University Press, 2007, pp. 11–12.

45 Resende-Santos, "Anarchy and the Emulation of Military Systems: Military Organization and Technology in South America, 1870–1930", p. 200.

46 Regarding the internal-external balancing distinction, Waltz states: "Internal efforts (moves to increase economic capability, to increase military strength, to develop clever strategies) and external efforts (moves to strengthen and enlarge one's own alliance or to weaken and shrink an opposing one)." See Waltz, *Theory of International Politics*, p. 118.

47 Waltz, *Theory of International Politics*, p. 97.

48 See about polarity and distribution of power, William R. Thompson, "Polarity, the Long Cycle, and Global Power Warfare", *The Journal of Conflict Resolution*, 30:4, 1986, pp. 587–615. Manus I. Midlarsky and Ted Hopf, "Polarity and International Stability", *The American Political Science Review*, 87:1, 1993, pp. 171–180. Edward D. Mansfield, "Concentration, Polarity, and the Distribution of Power", *International Studies Quarterly*, 37:1, 1993, pp. 105–128. Gregory A. Raymond and Charles W. Kegley Jr, "Polarity, Polarization, and the

Transformation of Alliance Norms", *The Western Political Quarterly*, 43:1, 1990, pp. 9–10. Nuno P. Monteiro, *Theory of Unipolar Politics*, Cambridge University Press, 2014.
49 Waltz, *Theory of International Politics*, pp. 161–170.
50 Resende-Santos, "Anarchy and the Emulation of Military Systems: Military Organization and Technology in South America, 1870–1930", pp. 201–203.
51 Kenneth N. Waltz, "The Origins of War in Neorealist Theory", *The Journal of Interdisciplinary History*, 18:4, 1988, pp. 620–624.
52 Resende-Santos, "Anarchy and the Emulation of Military Systems: Military Organization and Technology in South America, 1870–1930", p. 202.
53 Goldman and Andres, "Systemic Effects of Military Innovation and Diffusion", p. 83.
54 Resende-Santos, "Anarchy and the Emulation of Military Systems: Military Organization and Technology in South America, 1870–1930", p. 213.
55 Benjamin Frankel, "The Brooding Shadow: Systemic Incentives and Nuclear Weapons Proliferation", *Security Studies*, 2:3–4, 1993, p. 37.
56 Bruce Bueno de Mesquita, "Systemic Polarization and the Occurrence and Duration of War", *Journal of Conflict Resolution*, 22:2, 1978, p. 247.
57 Waltz, *Theory of International Politics*, pp. 134–137.
58 Waltz, *Theory of International Politics*, pp. 169–170.
59 Waltz, *Theory of International Politics*, pp. 170–172.
60 Kenneth N. Waltz, "Structural Realism After the Cold War", *International Security*, 25:1, 2000, p. 28.
61 Theo Farrell, "The Dynamics of British Military Transformation", *International Affairs*, 84:4, 2008, p. 780.
62 Farrell, "The Dynamics of British Military Transformation", p. 780.
63 Can Kasapoğlu and Barış Kırdemir, "The Rising Drone Power: Turkey On The Eve Of Its Military Breakthrough", https://www.jstor.org/stable/resrep21043?seq=1#metadata_info_tab_contents, accessed on 17.05.2021.
64 Everett M. Rogers, *Diffusion of Innovations*, 3rd Edition, The Free Press, New York, 1983, p. 176.
65 Ronald E. Rice and Everett M. Rogers, "Reinvention in the Innovation Process", *Knowledge: Creation, Diffusion, Utilization*, 1:4, 1980, p. 501.
66 Rogers, *Diffusion of Innovations*, p. 146.
67 Barry Buzan, Charles Jones and Richard Little, *The Logic of Anarchy: Neorealism to Structural Realism*, Columbia University Press, New York, 1993, p. 40.
68 Mike Eckel, "Drone Wars: In Nagorno-Karabakh, The Future Of Warfare Is Now", https://www.rferl.org/a/drone-wars-in-nagorno-karabakh-the-future-of-warfare-is-now/30885007.html, accessed on 17.05.2021. Nicholas Velazquez, "Rise of a "Drone Superpower?" Turkish Drones Upending Russia's Near Abroad", https://www.geopoliticalmonitor.com/turkish-drones/, accessed on 20.06.2021. "After Big Wins, Interest in Turkish Combat Drones Soars", https://www.france24.com/en/live-news/20210319-after-big-wins-interest-in-turkish-combat-drones-soars., accessed on 17.05.2021.
69 Francis Fukuyama, "Droning On in the Middle East".
70 Barton J. Bernstein, "The Cuban Missile Crisis: Trading the Jupiters in Turkey?", *Political Science Quarterly*, 95:1, 1980, pp. 98–100.
71 Nur Bilge Criss, "Strategic Nuclear Missiles in Turkey: The Jupiter affair, 1959–1963", *The Journal of Strategic Studies*, 20:3, 1997, pp. 115–120.
72 Quincy Wright, "Legal Aspects of the U-2 Incident", *The American Journal of International Law*, 54:4, 1960, pp. 836–854.
73 Lyndon B. Johnson and Ismet Inonu, "President Johnson and Prime Minister Inonu: Correspondence between President Johnson and Prime Minister Inonu, June 1964, as Released by the White House, January 15, 1966", *Middle East Journal*, 20:3, 1966, pp. 386–393.

74 William Hale, *Turkish Foreign Policy since 1774*, 3rd Ed., Routledge, New York, 2013, pp. 112–117.

75 Sibel Düz, "The Ascension of Turkey as a Drone Power: History, Strategy, and Geopolitical Implications", https://www.setav.org/en/analysis-the-ascension-of-turkey-as-a-drone-power-history-strategy-and-geopolitical-implications/., accessed on 21.06.2021.

76 "Poland to Become First NATO Country to Buy Turkish Drones", https://www.reuters.com/world/europe/poland-become-first-nato-country-buy-turkish-drones-2021-05-22/., accessed on 20.06.2021.

77 Düz, "The Ascension of Turkey as a Drone Power: History, Strategy, and Geopolitical Implications", pp. 8–9.

78 "Syria War: Turkish Forces 'Push into IS-held al-Bab'", https://www.bbc.com/news/world-middle-east-38916836., accessed on 16.05.2021. Fabrice Balanche, "Syria conflict: What is at stake in the battle of al-Bab?", https://www.bbc.com/news/world-middle-east-38939492., accessed on 16.05.2021.

79 Mevlut Cavusoglu, "The Meaning of Operation Olive Branch", https://foreignpolicy.com/2018/04/05/the-meaning-of-operation-olive-branch/., accessed on 16.05.2021.

80 Arda Mevlütoğlu, "Bahar Kalkanı Harekâtı ve Milli Muharip Uçak Projesi İçin Bazı Dersler", https://m5dergi.com/son-sayi/bahar-kalkani-harekati-ve-milli-muharip-ucak-projesi-icin-bazi-dersler/., accessed on 20.05.2021.

81 Ridvan Bari Urcosta, "The Revolution in Drone Warfare: The Lessons from the Idlib De-Escalation Zone", https://www.airuniversity.af.edu/JEMEAA/Article-Display/Article/2329510/the-revolution-in-drone-warfare-the-lessons-from-the-idlib-de-escalation-zone/., accessed on 20.06.2021.

82 Alexander Gabuev, "Viewpoint: Russia and Turkey – Unlikely Victors of Karabakh Conflict", https://www.bbc.com/news/world-europe-54903869., accessed on 10.06.2021.

83 "Azerbaijani President Ilham Aliyev: 'We Never Deliberately Attacked Civilians'", https://www.france24.com/en/asia-pacific/20201014-azerbaijani-president-ilham-aliyev-we-never-deliberately-attacked-civilians., accessed on 10.06.2021.

84 Henry Foy, "Drones and Missiles Tilt War with Armenia in Azerbaijan's Favour", https://www.ft.com/content/6acddc7d-cda5-44b2-9e5c-f6863f7bb9e7., accessed on 10.06.2021.

85 Jason Pack and Wolfgang Pusztai, "Turning the Tide: How Turkey Won the War for Tripoli",https://www.mei.edu/sites/default/files/2020-11/Turning%20the%20Tide%20-%20How%20Turkey%20Won%20the%20War%20for%20Tripoli.pdf., accessed on 11.06.2021. Alex Gatopoulos, "'Largest drone war in the world': How airpower saved Tripoli", https://www.aljazeera.com/news/2020/5/28/largest-drone-war-in-the-world-how-airpower-saved-tripoli., accessed on 11.06.2021.

86 Will Smith, "Turkey's Drones and Proxies are Turning the Tide of War", https://nationalinterest.org/feature/turkey%E2%80%99s-drones-and-proxies-are-turning-tide-war-194576., accessed on 12.10.2021.

References

"After Big Wins, Interest in Turkish Combat Drones Soars", https://www.france24.com/en/live-news/20210319-after-big-wins-interest-in-turkish-combat-drones-soars, accessed on 17.05.2021.

"Azerbaijani President Ilham Aliyev: 'We Never Deliberately Attacked Civilians'", https://www.france24.com/en/asia-pacific/20201014-azerbaijani-president-ilham-aliyev-we-never-deliberately-attacked-civilians, accessed on 10.06.2021.

Balanche, Fabrice, "Syria Conflict: What is at Stake in the Battle of al-Bab?" https://www.bbc.com/news/world-middle-east-38939492, accessed on 16.05.2021.

Bernstein, Barton J., "The Cuban Missile Crisis: Trading the Jupiters in Turkey?", *Political Science Quarterly*, 95:1, 1980, pp. 97–125.

Buzan, Barry, Jones, Charles and Little, Richard, *The Logic of Anarchy: Neorealism to Structural Realism*, Columbia University Press, New York, 1993.

Cavusoglu, Mevlut, "The Meaning of Operation Olive Branch", https://foreignpolicy.com/2018/04/05/the-meaning-of-operation-olive-branch/., accessed on 16.05.2021.

Cirincione, Joseph, *Bomb Scare: The History and Future of Nuclear Weapons*, Columbia University Press, New York, 2007.

Criss, Nur Bilge, "Strategic Nuclear Missiles in Turkey: The Jupiter affair, 1959–1963", *The Journal of Strategic Studies*, 20:3, 1997, pp. 97–122.

De Mesquita, Bruce Bueno, "Systemic Polarization and the Occurrence and Duration of War", *Journal of Conflict Resolution*, 22:2, 1978, pp. 241–267.

Durkheim, Emile, *The Rules of Sociological Method*, (trs.) W.D. Halls, New York, The Free Press, 1982.

Düz, Sibel, "The Ascension of Turkey as a Drone Power: History, Strategy, and Geopolitical Implications", https://www.setav.org/en/analysis-the-ascension-of-turkey-as-a-drone-power-history-strategy-and-geopolitical-implications/, accessed on 21.06.2021.

Eckel, Mike, "Drone Wars: In Nagorno-Karabakh, The Future of Warfare Is Now", https://www.rferl.org/a/drone-wars-in-nagorno-karabakh-the-future-of-warfare-is-now/30885007.html, accessed on 17.05.2021.

Farrell, Theo, "The Dynamics of British Military Transformation", *International Affairs*, 84:4, 2008, pp. 777–807.

Fontenoy, Paul, *Aircraft Carriers: An Illustrated History of Their Impact*, Oxford, ABC-CLIO, 2006.

Foy, Henry, "Drones and missiles tilt war with Armenia in Azerbaijan's favour", https://www.ft.com/content/6acddc7d-cda5-44b2-9e5c-f6863f7bb9e7, accessed on 10.05.2021.

Frankel, Benjamin, "The Brooding Shadow: Systemic Incentives and Nuclear Weapons Proliferation", *Security Studies*, 2:3–4, 1993, pp. 37–78.

Fukuyama, Francis, "Droning on in the Middle East", https://www.americanpurpose.com/blog/fukuyama/droning-on/, accessed on 10.05.2021.

Gabuev, Alexander, "Viewpoint: Russia and Turkey – Unlikely Victors of Karabakh Conflict", https://www.bbc.com/news/world-europe-54903869, accessed on 10.06.2021.

Gatopoulos, Alex, "'Largest Drone War in the World': How Airpower saved Tripoli", https://www.aljazeera.com/news/2020/5/28/largest-drone-war-in-the-world-how-airpower-saved-tripoli, accessed on 11.06.2021.

Goldman, Emily O. and Andres, Richard B., "Systemic Effects of Military Innovation and Diffusion", *Security Studies*, 8:4, 1999, pp. 79–125.

Goldsmith, Benjamin, "Imitation in International Relations: Analogies, Vicarious Learning, and Foreign Policy", *International Interactions*, 29:3, 2003, pp. 237–267.

Gutfeld, Arnon, "From 'Star Wars' to 'Iron Dome': US Support of Israel's Missile Defense Systems", *Middle Eastern Studies*, 53:6, 2017, pp. 934–948.

Guvenc, Serhat and Uyar, Mesut, "Lost in Translation or Transformation? The Impact of American Aid on the Turkish Military, 1947–60", *Cold War History*, 22:1, 2022, pp. 59–77.

Hale, William, *Turkish Foreign Policy since 1774*, 3rd Edition, Routledge, New York, 2013.

Harrison, Ewan, *The Post-Cold War International System: Strategies, Institutions and Reflexivity*, Routledge, New York, 2004.

Hoffmann, Stanley, "An American Social Science: International Relations", *Daedalus*, 106:3, 1977, pp. 41–60.

Holsti, K. J., *International Politics: A Framework for Analysis* 7th Edition, Prentice-Hall International Editions, New Jersey, 1995.

Ikenberry, G. John and Kupchan, Charles A., "Socialization and Hegemonic Power", *International Organization*, 44:3, 1990, pp. 283–315.

Jervis, Robert, *System Effects: Complexity in Political and Social Life*, Princeton University Press, Princeton, 1997.

Johnson, Lyndon B. and Inonu, Ismet, "President Johnson and Prime Minister Inonu: Correspondence between President Johnson and Prime Minister Inonu, June 1964, Release by the White House, Jan 15, 1966", *Middle East Journal*, 20:3, 1966, pp. 386–393.

Kasapoğlu, Can and Kırdemir, Barış, "The Rising Drone Power: Turkey on the Eve of Its Military Breakthrough", https://www.jstor.org/stable/resrep21043?seq=1#metadata_info_tab_contents, accessed on 17.05.2021.

Lynn, John A., "The Evolution of Army Style in the Modern West, 800-2000", *The International History Review*, 18:3, 1996, pp. 505–545.

Mevlütoğlu, Arda, "Bahar Kalkanı Harekatı ve Milli Muharip Uçak Projesi İçin Bazı Dersler", https://m5dergi.com/son-sayi/bahar-kalkani-harekati-ve-milli-muharip-ucak-projesi-icin-bazi-dersler/, accessed on 20.05.2021.

Onuf, Nicholas, "Structure? What Structure?", *International Relations*, 23:2, 2009, pp. 183–199.

Ortaylı, İlber, *İmparatorluğun En Uzun Yüzyılı*, 25th Edition, Alkım, İstanbul, 2006.

Ortaylı, İlber, *Osmanlı İmparatorluğu'nda Alman Nüfuzu*, 11th Edition, Timaş, İstanbul, 2010.

Pack, Jason and Pusztai, Wolfgang, "Turning the Tide: How Turkey Won the War for Tripoli", https://www.mei.edu/sites/default/files/2020-11/Turning%20the%20Tide%20-%20How%20Turkey%20Won%20the%20War%20for%20Tripoli.pdf, accessed on 11.06.2021.

"Poland to Become First NATO Country to Buy Turkish Drones", https://www.reuters.com/world/europe/poland-become-first-nato-country-buy-turkish-drones-2021-05-22/, accessed on 20.06.2021.

Posen, Barry R., "Nationalism, the Mass Army, and Military Power", *International Security*, 18:2, 1993, pp. 80–124.

Resende-Santos, Joao, "Anarchy and the Emulation of Military Systems: Military Organization and Technology in South America, 1870–1930", *Security Studies*, 5:3, 1996, pp. 193–260.

Resende-Santos, Joao, *Neorealism, States, and the Modern Mass Army*, Cambridge University Press, Cambridge, 2007.

Rice, Ronald E. and Rogers, Everett M., "Reinvention in the Innovation Process", *Knowledge: Creation, Diffusion, Utilization*, 1:4, 1980, pp. 499–514.

Rogers, Everett M., *Diffusion of Innovations*, 3rd Edition, The Free Press, New York, 1983.

Schimmelfennig, Frank, "International Socialization in the New Europe: Rational Action in an Institutional Environment", *European Journal of International Relations*, 6:1, 2000, pp. 109–139.

Singer, J. David, "The Level-of-Analysis Problem in International Relations", *World Politics* (The International System: Theoretical Essays), 14:1, 1961, pp. 77–92.

Smith, Will, "Turkey's Drones and Proxies are Turning the Tide of War", https://nationalinterest.org/feature/turkey%E2%80%99s-drones-and-proxies-are-turning-tide-war-194576, accessed on 12.10.2021.

"Syria War: Turkish Forces 'Push into IS-held al-Bab'", https://www.bbc.com/news/world-middle-east-38916836, accessed on 16.05.2021.

Tang, Shiping, "Fear in International Politics: Two Positions", *International Studies Review*, 10:3, 2008, pp. 451–471.

Thies, Cameron G., "State Socialization and Structural Realism", *Security Studies*, 19:4, 2010, pp. 689–717.

Urcosta, Ridvan Bari, "The Revolution in Drone Warfare: The Lessons from the Idlib De-Escalation Zone", https://www.airuniversity.af.edu/JEMEAA/Article-Display/Article/2329510/the-revolution-in-drone-warfare-the-lessons-from-the-idlib-de-escalation-zone/, accessed on 20.06.2021.

Uyar, Mesut and Erickson, Edward J., *A Military History of the Ottomans: From Osman to Atatürk*, ABC-CLIO, Santa Barbara, 2009.

Velazquez, Nicholas, "Rise of a "Drone Superpower?" Turkish Drones Upending Russia's Near Abroad", https://www.geopoliticalmonitor.com/turkish-drones/, accessed on 20.06.2021.

Waltz, Kenneth, *Man, The State and War: A Theoretical Analysis*, Columbia University Press, New York, 1959.

Waltz, Kenneth, *Theory of International Politics*, Addison-Wesley, Massachusetts, 1979.

Waltz, Kenneth N., "The Origins of War in Neorealist Theory", *The Journal of Interdisciplinary History*, 18:4, 1988, pp. 615–628.

Waltz, Kenneth N., "Structural Realism After the Cold War", *International Security*, 25:1, 2000, pp. 5–41.

Wright, Quincy, "Legal Aspects of the U-2 Incident", *The American Journal of International Law*, 54:4, 1960, pp. 836–854.

6 Military Innovation in the Context of Irregular Warfare

The British, American, and Turkish Militaries

Emrah Özdemir

Introduction

Clausewitz relates war to a chameleon that can adapt to its surroundings and argues that war must also be able to adapt to continuously changing circumstances.[1] The best course of action, the best structure, and the best strategy may vary in line with different variables, while being able to adapt to transformations as soon as possible is vital due to the changing character of war, which is shaped by the influence of political, economic, and social transformations. In other words, the actors of international relations and the changes in their policies also bring about change in war.[2] For example, the advancement of steam technology in the Industrial Age led to increased production as well as wars between larger armies with more intense firepower. Unable to withstand the firepower, infantry collapsed into trenches, marking the beginning of an era in which trench warfare was adopted instead of offense and maneuver. This trench-war logic and defensive perspective continued until the Germans implemented the Blitzkrieg doctrine in World War II. The Cold War's nuclear deterrence also resulted in the development of new strategic views, with the conventional lessons from World War II beginning to be seen as superfluous after 1945. As another example, nation-states' monopoly on war has decreased since 1990 with the emergence of non-state armed groups.

Therefore, change in warfare sometimes occurs with the emergence of new technologies and sometimes through alterations in the political environment. This change also affects armies' structures, weapons, equipment, and strategies. Using the term "innovation" has recently come to be preferred when explaining these processes. However, the concept of innovation not only means a technological invention or development of a new weapon system in military terms, but can also be expressed as new approaches and ideas for keeping up with political, social, cultural, and administrative changes in the broadest sense. For this reason, describing only a developing technology as an innovation would be an improper and incomplete statement. The ability to adapt in the face of developing situations and to reveal a new doctrine or weapon system includes using existing structures, weapons, and equipment differently and more creatively than in the usual methods.

DOI: 10.4324/9781003327127-7

From this perspective, the use of irregular warfare in situations where conventional warfare is ineffective at achieving victory against an adversary with greater conventional capability may also be deemed innovative. However, as a strategy that the weak use against the powerful, irregular warfare cannot be seen as an innovative strategy in all circumstances. One essential principle for evaluating a new weapon system or process as an innovation is that it must cause a radical change in the military mindset and military culture. In this sense, the primary purpose of this chapter is to analyze cases in which irregular warfare has created a fundamental change in military thinking and culture by using an innovative approach.

The chapter will first make a conceptual introduction to resolve the confusion created around the concept of irregular warfare. This section will examine the meaning of irregular warfare and the types of operations. Next, it will examine three historical examples where irregular warfare has been innovatively and creatively used. The first of these examples is the British experience behind German lines during World War II. The second involves the practices outside conventional warfare that the US Army attempted to introduce during the Vietnam War. The last example includes the counterinsurgency operations of the coalition forces in Afghanistan and Iraq after the September 11 attacks. After examining these cases, the chapter will focus on the meaning of the concept of contemporary irregular warfare and how it is interpreted and will explore the changing irregular warfare approach of the Turkish Armed Forces. The main argument of the chapter is that irregular warfare has been evaluated as a novel and innovative option in times of stalemate, crisis, and despair.

The Concept of Irregular Warfare in the Context of Innovation and Institutional Change

Although irregular warfare is a rather old concept, it did not start occupying a broad place in the literature until the post-Cold War period, especially after the September 11 attacks. Today, armed conflicts continue to arise mostly between non-state entities and states; this inclination often causes irregular warfare to be the dominant strategy. Irregular warfare is a comprehensive term that covers all military activities falling outside those of the conventional warfare waged by regular army units. When conventional warfare is defined as the use of organized violence for political goals by national or conventional armies on a battlefield, the essence of the wars experienced in the pre-modern era can be claimed to have been irregular warfare. In this regard, the fight of an agriculturally established society in ancient times to defend its possessions against invaders may also be seen as a type of irregular warfare. According to this perspective, irregular warfare is the earliest and most fundamental form of combat.

The story of David and Goliath, one of the oldest known examples of irregular warfare, is proof of how deep-rooted the concept is. Goliath, being physically much stronger than an average person in the well-known story, was also equipped with a shield, armor, and sword. In the face of such a superior and invincible power, coming out with a conventional and predictable method would have been considered as accepting defeat from the beginning. Thus, David brought a different perspective to the physical struggle by focusing on the superior skills he had that Goliath did not. First, David had characteristics such as agility and speed that came to the fore, but adding the slingshot as a weapon Goliath never expected was the key to a historic victory.[3] Although both the physical characteristics of David and the slingshot were already present elements, their use in combination, from a different point of view and against a superior power like Goliath could be considered as an innovative practice for that period.

Irregular warfare is a standard definition of military operations other than traditional, conventional strategies and tactics. Although many sources have attempted to rename certain past experiences as irregular warfare, the defining characteristic of the concept is that it is an alternative to traditional warfare conducted by regular troops. While generally accepted as a method the weak use against the strong, cases also show that regular armies have also sometimes used it to gain superiority over an enemy. Considering the theoretical definition of the concept, despite the French having the first systematic experience defined as guerrilla war in their Spanish campaign, and the general belief being that the first example of successful counterinsurgency was conducted by the British in Malaya, the definition of the concept of irregular warfare is based on US military doctrine.

The ambiguity and fluidity of the concept have led to different perspectives and definitions, even in US military literature. For this reason, although different definitions are found in various official documents, irregular warfare is generally defined as a struggle based on violence between state and non-state actors to gain legitimacy in the mind of the population. Another dimension emphasized in the definitions is that it is managed with an indirect and asymmetric approach.[4] Even this comprehensive definition is unable to place the concept within a clear framework, and it does contain two essential typologies. The first is the population-centric warfare approach that should be carried out in order to gain legitimacy and positively affect the people. According to this approach, the aim is not just to eliminate the enemy or neutralize insurgents. Furthermore, it envisages a comprehensive effort as a permanent and sustainable solution for creating a socio-cultural, economic, and political structure that will disallow the re-emergence of the insurgency. The other typology is associated with the concept of low-intensity conflict, which characterizes the operations carried out behind enemy lines or within other countries' territory.[5]

A more specific definition of the concept of irregular warfare can be found in the US Army's Special Operations manual. With a similar point of view, this definition describes the concept as a general term that includes operations other than conventional battles, but it also lists the types of operations. These types of operations are: (1) unconventional warfare (UW), which is defined as "organizing and supporting the insurgency and resistance movements against an enemy state"; (2) the support and execution of security activities against friendly and allied states, known as foreign internal defense [FID]; (3) counterinsurgency [COIN]; (4) counterterrorism (CT); and (5) stability operations (Stability OPs).[6]

The most striking among these types of operations is unconventional warfare. Irregular warfare (IW) and unconventional warfare (UW) have been used as synonyms in the literature, which has created confusion. As a form of irregular warfare, UW includes all armed and unarmed activities against the rule of an adversary state to overthrow it and replace it with new governance more in harmony with national policies. Intelligence agencies and special forces have carried out such operations in countries such as Nicaragua, the Philippines, Indonesia, Cuba, Chile, and the Congo during the Cold War. In such a period when the fear of nuclear war, which was able to affect the whole world, restricted the use of conventional capacity, irregular warfare was considered an innovative approach, albeit difficult to approve. These definitional complexities will be more clarified through selected cases in the historical process.

Irregular Warfare Experiences of World War II

Pursuing the idea of a great and mighty empire, Germany had unexpected success in the initial phase of World War II (WWII) between 1939 and 1942. The underlying reason for Germany's significant military success within such a short time was their ability to use the contemporary military technology of the period in the most effective way. The tank brought back to the battlefield the infantry who had remained stuck in trenches in World War I (WWI). Its firepower and maneuverability again had the potential to allow the fate of a war to be determined. While the tank did not make much sense to the post-WWI British and French policymakers, the German army, on the other hand, was able to see the potential for this fast-moving war machine with armor and superior firepower to play a leading role in the wars of the future. The main reason for the tank's success lies in the doctrine that was developed regarding how this technology could be used in conjunction with other existing weapon systems and military structures, along with its normal development and use. Although the tank technology of the French army initially exceeded that of the Germans in this period, the French had interpreted the role of the tank as a support weapon, similar to artillery. As a result of the superior German Blitzkrieg doctrine, the French army surrendered, and the British army were forced to withdraw 300,000 soldiers from continental Europe under very dramatic conditions. In such a problematic situation, Britain was now seen as the German's new target and did not have many options. Winston Churchill needed time to recover the British army and find allies against the Germans. The time that was critically required could perhaps be obtained through the unconventional means of demoralizing and distracting the immensely strong German army.

The colonial experiences of the British army and bureaucracy in retaining their colonies enabled them to gain profound experience with irregular warfare. For example, British officers such as Orde Wingate and Dudley Clarke in Palestine saw how successful the guerrilla struggles were against the regular armies and followed and applied similar methods. Clarke contributed to forming the first examples established for operating behind German lines, such as the Norwegian volunteers and the 5th Scottish Guards Battalion against the Russians in Finland.[7]

These examples can be described as relatively classical approaches. However, the real innovative solution is shown in the Special Operations Executive (SOE) established by Churchill. Although this structure follows classic commando or special unit tactics, it was a very creative approach in terms of organization and personnel resources. Members of the SOE were selected from civilians who voluntarily underwent special training. Most of the personnel were selected from volunteers who had come to Britain from German-occupied areas and been assigned to support and organize resistance movements in these areas. These personnel were able to perform activities such as information gathering, reconnaissance, and sabotage without drawing attention behind enemy lines. Beyond providing a significant distraction by creating pressure on the German army, the structure and personnel also benefited by gathering serious intelligence.[8]

Other important irregular warfare units that the British actively used were the Long-Range Desert Groups (LRDG). These units, operating in north Africa, were founded by Major Ralph A. Bagnold. LRDG carried out over 200 successful operations in a five-year period. In addition to activities such as reconnaissance, surveillance, raids, and sabotage behind enemy lines, they were also active in providing contact and supply activities between troops. The Special Air Service (SAS) and Popski's Private Army were other irregular warfare units of the British Army that also carried out very successful operations.

By developing the operative level and using the concept of air-ground joint operations effectively through the efficient use of air forces and armored units, the Germans had enormous conventional warfare capabilities at the beginning. By aiming to gain time and deliver as much physical damage and loss of morale as possible to the Germans until the Allies attained the same type of conventional capacity, these special units and their irregular war tactics and strategies brought success beyond expectations. Operations that resulted in casualties and failures due to an initial lack of experience became quite effective over time. Successful operations carried out by these special units also had a significant effect on the morale of both the soldiers on the frontlines and the people in the homeland.

The Experiences of the United States in Vietnam

US intervention followed the French withdrawing from Vietnam after their defeat in a long-lasting guerrilla campaign. The French had attempted to win through conventional means and paid a high price in terms of casualties due to the irregular war North Vietnam and the Viet Cong conducted under the political leadership of Ho Chi Minh and the military leadership of Vo Nguyen Giap. Although the USA had this example and the US Military Assistance Advisory Group (MAAG) had witnessed the French defeat, the USA still chose to support South Vietnam with a conventional military understanding in line with its own political interests, through the self-confidence effect provided by nuclear weapons. Thus, the US military's course of action in Vietnam was based on air and sea bombardments as well as the activities of the advisory and aid teams with regard to rebuilding the South Vietnamese army.

After the 1954 Geneva Peace Treaty, an anti-communist Republic of Vietnam was established in South Vietnam under the leadership of Ngo Dinh Diem. Although the public did not approve of Diem, the USA supported the government politically and militarily due to the nature of the Cold War. The military aim of the USA was to establish a regular army structure against threats to South Vietnam, especially from North Vietnam, and to support it with military equipment and weapon systems.[9] The insufficient firepower and failure of the activities of the regular units of the South Vietnamese army necessitated the US Army's intervention with ground troops. The US Army's approach followed the conventional tradition and adopted the tactics of search and destroy with an enemy-centric perspective.[10]

The conviction that a conventional approach alone would not be effective yielded the testing of an unconventional population-centric strategy during the administration of US President John F. Kennedy. This approach to counterinsurgency was the product of an innovative perspective compared to the enemy-centric conventional approach. Although previously inspired by French and British experiences, this effort involved different institutions such as the Department of Defense, the Department of State, and the CIA and was based on eliminating the people's dissatisfaction and ensuring their isolation from the North Vietnamese administration and the Viet Cong. This altered approach led to a doctrinal transformation in the US Army that successfully entered military literature. The *US FM 100-5 Operations Manual* attributed the insurgency movements in the country to the inadequacy of infrastructure, the lack of health and education services, and eventually the lack of a stable state system able to provide these services. In this regard, a military manual envisioned applying a civil and political solution to a complex problem far beyond the capability of the military. As stated by Galula, American soldiers in Vietnam had to take responsibility as workers, engineers, teachers, and health workers, albeit outside their field of expertise, in order to win the people over.[11] The idea was that they would only be able to ensure security in this way. Presented as the only alternative to the unsuccessful conventional approach, this population-centric comprehensive approach was unfortunately again replaced by the approach of the Johnson administration that was based on intense firepower and maneuver warfare.

Although the USA was unable to achieve its expected outcome, the use of military units outside of their firepower and combat capacity to neutralize the enemy can be considered a very remarkable attitude. The approach of winning hearts and minds, which was described as a population-centric point of view in insurgency and guerrilla movements, is a perspective that had previously been tried by the British in the Boers and Malaya cases and the French in Indochina and Algeria. However, this was a highly innovative approach for the US military, which had taken a deep-rooted conventional stance in Vietnam. Agroville, Strategic Hamlet, and Civil Operations and Revolutionary Development Support – CORDS – are examples of the applications that were included in this new strategy.

With these programs, non-kinetic activities aimed at winning hearts and minds were placed under the responsibility of a civilian administrator at the ambassador level, and an ambassador was assigned for the first time under the command of a soldier. Meanwhile, the military structures that would carry out these activities were placed under the administration of a civilian assigned in the field.[12] Moreover, the program was designed as a joint effort of different agencies such as the Central Intelligence Agency (CIA), Department of Defense, the United States Information and Communications Agency, and the US Agency for International Development (USAID) and considered evidence that efforts to counter such insurgency required a comprehensive approach.[13]

The primary distinction between the war in Vietnam fought by the US Army, and conventional warfare was the lack of a clearly determined frontline. Consequently, the lack of frontlines in such a war resulted in the concepts of safe zones and base areas gaining importance. This situation caused the concept of operating behind enemy lines, one of the most important features of irregular warfare, to become valid for the entire US Army. Despite this, the US military cannot easily be said to have readily adapted to this situation. However, the US Army carried out significant innovative applications focused on firepower and maneuverability, the most important of which was the use of helicopters.

Helicopter technology was first used to evacuate casualties during the Korean War and greatly affected the operation in Vietnam and its aftermath.[14] Utility helicopters such as the Sheridan with their capacity to carry tanks and armored vehicles, and the UH-1 known as the Huey with a lesser capacity were used to reinforce base areas, provide logistical support, and ensure the safe transport of military units. In addition to these purposes, they were also very effective in controlling the escape routes of the North Vietnamese troops and Viet Cong guerrillas.[15] The AH-1 Cobra helicopter provided effective fire support to the infantry units and was accepted as the pioneer of attack helicopter technology. Again, the UH-1 helicopter, installed with the AN/VSS-3 Xenon white and infrared light system, AN/TVS-4-night vision system, and XM-27 EI 7.62 mm minigun multi-barreled machine gun system, provided significant advantages, especially in open areas at night. In this way, these helicopters limited the night advantage the enemy had previously used to take the initiative.[16] However, although progress was made in areas such as reconnaissance, surveillance, and human and signal intelligence, special troops were also included in the operation, and the US military's conservative approach to conventional warfare was largely preserved.

Military Innovation During the Iraq and Afghanistan Wars and Beyond

The Vietnam experience illustrates that the technologies and approaches developed in the long-term struggle were unable to overcome the rigid limitations of conventional military culture, resisting institutional change. This cultural conservatism about military change caused the US military to be skeptical of irregular warfare approaches for a long time and to ignore their importance.

On the other hand, with the end of the Cold War, US military power was maximized in terms of firepower, speed, maneuverability, and decision-making capacity with technological means based on the approach of revolution in military affairs (RMA). This approach caused military and political authorities of the USA to think they were undefeatable. The USA strongly reinforced this idea of conventional superiority during the Iraq invasion of Kuwait. Believing that it would not lose any conventional wars, the USA became the Goliath of the modern age, and this overconfidence left it vulnerable against primitive methods such as terrorist attacks and irregular warfare. The reflexive intervention into Afghanistan after the terrorist attacks of September 11 created the idea that the Taliban administration had been expelled from the country and largely neutralized. However, the Taliban administration was able to adapt to this developing situation very quickly, had recovered by 2004 and started an insurgency movement that marked the beginning of a chaotic conflict in the country.[17]

The apparent superiority of coalition forces regarding conventional means made the irregular warfare approach the only option for the Taliban. The Taliban achieved a very advantageous situation with this strategy and knew that the longer they could prolong the conflict and struggle against this enemy within the country, the better their chances of success. Therefore, they aimed not only to achieve military victory but also to impose their capacity to undertake administration of the country. The so-called civil administration mechanisms were also established in the regions under their control.

The easy success in Afghanistan had encouraged the US political and military elites to perform another military intervention in Iraq. Saddam Hussein's administration in Iraq was seen as a threat to US national interests and its regional policies. The deceptive information that the Iraqi administration might have weapons of mass destruction (biological and nuclear weapons) legitimized the occupation of this country for the purpose of stability and peace. However, this baseless allegation pushed the USA and its coalition partners into a complex irregular warfare experience that they could not easily understand. Sunni resistance groups and subsequently Shiite resistance groups initiated a serious insurgency movement against the coalition.

Trying to avoid such conflict zones and struggles after Vietnam, the USA unexpectedly found itself in another counterinsurgency conflict and had to adapt itself to irregular warfare approaches. The coalition forces had had serious problems initially understanding and analyzing the threat they faced and accepted the population-centric approach of counterinsurgency as a strategy that could bring success. Efforts led by General David Petraeus in 2006 and the *FM 3-24 Counterinsurgency Manual* had a significant impact on this change. As a result of the efforts made by the low-ranking officers in Iraq, the population-centric counterinsurgency approach was adopted by the tactical commanders and used before *FM 3-24*'s publication without any coordination or unity of effort. Thus, the population-centric approach to opposing insurgency is not a top-down approach; on the contrary, it is the institutionalization of tactical methods adopted by the tactical-level unit commanders.[18] Winning the people was determined as the primary goal, rather than a kinetic approach targeting the enemy. At this point, the approach of winning hearts and minds, although not novel, was embraced once again as a remedy.

The most striking difference in this approach was the acceptance of the social sciences, especially social anthropology, as one of the main components of the operation at both the planning and implementation levels. In this regard, human terrain teams can be considered a remarkable innovation. The military decision-making mechanisms in Iraq and Afghanistan were thought to need social, cultural, economic, historical, and demographic information about the operation areas for these mechanisms to work effectively. Consequently, the teams included research experts such as anthropologists and sociologists in addition to the military personnel responsible for ensuring security. Although these mixed teams were abolished as of 2014, awareness of the cultural and sociological structure in planning and operations has become relatively permanent in the US Army.

Another critical approach involved provincial reconstruction teams (PRTs).[19] Similar to the Strategic Hamlet Program in Vietnam, the approach envisaged the creation of safe villages and towns in rural areas, especially where the Taliban threat was intense. The aim was to create a perception in these settlements through both the public and the political authorities that the fight against the insurgency would be successful and provide the people with the necessary support regarding education, health, and food. The strategic approach that was applied in the struggles of Iraq and Afghanistan became a comprehensive approach for countering insurgency and was adopted not only by the armed forces and the Department of Defense but also by other public institutions of the USA and NATO.[20] The soldiers, whose primary purpose was determined as the use of arms to achieve the determined political goals, endeavored to re-establish political authority to win the armed struggle.

These new insurgency operations also caused drastic changes in the weapon, ammunition, and military structures. The new enemies could not use tanks or howitzers, and so using these conventional weapon systems against them was also useless.[21] Western countries, especially the US military, developed a brigade-level task force approach to ensure this modularity and mobility. Although tanks and cannons maintained their effectiveness, more aircraft, helicopters, and uncrewed aerial vehicles started to be used extensively as fire support assets.[22]

Moreover, the measures developed against the improvised explosive devices (IEDs) irregular elements frequently used against conventional armies, had innovative features. In particular, the High Mobility, Multipurpose Wheeled Vehicle (HMMWV or Humvee), which the US military had used since 1985, was vulnerable to IED attacks and had disadvantages such as its limited mobility and weight. In 2008, the search began for a modular platform to replace these vehicles with one that had increased mobility, was able to move in all kinds of terrains, and provided protection against IEDs. As a result of this effort, the Oshkosh M-ATV was developed; a mine resistant ambush protected (MRAP) all-terrain vehicle (ATV).

ISR (Intelligence, Surveillance, and Reconnaissance) technologies are other critical technological developments these experiences provided. Thanks to these high technologies, different platforms such as satellite, aircraft, and uncrewed aerial vehicles became integrated with sensor technologies, allowing easy exploration and detection. The communications and network logic UAV systems such as the RQ-1/MQ-1 Predator, MQ-9 Reaper, and RQ-4 Global Hawk utilized with units in the field have also provided the opportunity for quick intervention.[23] The experience gained in Iraq and Afghanistan regarding ISR technologies undeniably formed the basis of the observed development in uncrewed aerial vehicles.[24] In addition, smart missile systems as well as sensor and imaging technologies enabled more accurate decisions regarding target detection by reducing civilian casualties and collateral damage.[25]

With their reconnaissance and assault capacities, UAVs have become an inevitable part of the security architecture in the contemporary security environment. Countries such as Nigeria, Iraq, Türkiye, Saudi Arabia, and the United Arab Emirates fighting against insurgencies and terror groups have used UAVs effectively to combat terrorism and insurgency.[26] Even insurgents and terrorist groups seek to use UAVs and operate commercial drones by loading them with explosive materials and basic electronic systems. This necessity has created a massive market, and China is considered the current pioneer of UAV technology in terms of commercial capacity.[27] China has very effective and economic weapon systems and vehicles. For instance, between 2011 and 2019, 18 countries bought UAVs, 11 of whom (including Iraq and Jordan) obtained them from China.[28]

In the cases of Iraq and Afghanistan, although the population-centric approaches to counter insurgency achieved short-term success, they were insufficient at ensuring long-term peace and stability in both countries. These were unsustainable efforts whose human resource and economic costs exceeded what Western states could afford, especially the USA. Thus, these comprehensive state-building activities were replaced by more limited operations. In this context, the USA gave these same responsibilities to the PKK-affiliated SDF/YPG, an organization both the EU and the USA have designated as terrorist, while ensuring security was transferred to local security forces in Afghanistan and Iraq. In Syria and Iraq, the US government has supported terrorist organization PKK-affiliated establishments in accordance with its policies. In official documents and speeches, US officials have attempted to convince international society that YPG is an independent organization trying to stop ISIS. Self and Ferris[29] stated that PKK used this contingent policy as a shell game through sister organizations in Iran, Iraq, and Syria to reach the long-term objective of freeing Kurdistan. After 2014, population-centric irregular warfare activities turned into local elements being used through advice, training and weapons, equipment, and intelligence support. In this way, economic and political risks were minimized by keeping a limited number of soldiers in critical regions and on dangerous tasks.

Moreover, after the examples of Afghanistan and Iraq, private security and military companies have become a rising trend in conflict areas such as Syria, Libya, and Yemen. Different states such as the United Arab Emirates (UAE), Russia, and the USA use these private companies as their political and military assets on the ground, similar to local proxies. For example, the Russian private security company ChVK Wagner Group, owned by Yevgeny Prigozhin (who is a businessman with close ties to Russian President Putin), is alleged to be involved in the manipulation of elections and economic markets and to direct conflict in countries such as Syria, Sudan, and Libya. For example, the 50 ventilation devices, 10,000 coronavirus test kits, and 2,000 protective materials Russia sent to Syria in April 2020 were delivered by Wagner-linked Evro Polis, which also has a contract with a Syrian state-owned corporation.[30] This support can be evaluated as an effective activity for keeping the Assad regime in power, which means these companies can also be used in line with the countries' political and military interests.[31]

According to the UN report prepared by a panel of experts on Libya, ChVK Wagner was militarily active in Libya and maintained combat support for the Hafter Affiliated Groups (HAG). Between 2019 and 2020, ChVK Wagner carried out military tasks such as "artillery forward observation officers and forward air controllers, providing electronic counter-measures expertise and deploying as sniper teams".[32] In addition, Russian military aircraft have deployed this private company's personnel as well as deploying Russian Federation-maintained planes such as the Mig-29A, SU-24, and Pantsir S-1 missile systems for the company. Another private military and security intervention in Libya is Project Opus, which is conducted by three UAE-based companies (Lancaster 6 DMCC, L-6 FZE, and Opus Capital Asset Limited FZE). The project maintained military support for the HAG light and medium utility helicopters, armed assault rotary-wing aviation, intelligence, surveillance, and reconnaissance aircraft, light attack aircraft, and UAVs. These companies also deployed private military operatives to launch covered operations such as kidnapping or assassinating influential individuals and locating and destroying high-value targets.[33]

Providing public support is critical since these activities are carried out among the populace within urban areas. The most critical issue to consider is how to restrain military and economic capabilities in line with well-defined and limited political goals. Ensuring the establishment and training of local law enforcement agencies in the region is also essential based on the idea that military units alone will not be sufficient for re-establishing an environment of stability and peace in conflict zones. Supporting the armed forces' activities in these regions with the capabilities and capacities of the gendarmerie-type constabulary units makes a significant contribution. The application of these activities can be an example of NATO's increasingly important stability policing (SP) approach. SP is a concept that aims to replace or reinforce the local law-enforcement structure to maintain safety and public order in post-conflict zones.[34]

The Irregular Warfare Experiences and Innovative Solutions of the Turkish Military

Türkiye has inevitably had to intervene in conflicts occurring in the Middle East. To not intervene is impossible, due first to the refugee influx caused by the civil war in its neighbor Syria, and the civilian casualties caused by the rockets and mortar shells fired from the neighboring regions directly affecting Türkiye, and due second to the Syrian army's lack of control along the Türkiye-Syria border, after allowing terrorist organizations to dominate the region. Based on these reasons, Türkiye has carried out cross-border operations to create a safe zone along the border region and to neutralize the terrorist organizations in this area. These operations are Operation Euphrates Shield (August 24, 2016–March 29, 2017), Operation Olive Branch (January 20–March 24, 2018), Operation Peace Spring (October 9–17, 2019), and Operation Spring Shield (February 27–March 6, 2020) and follow irregular warfare and anti-terrorism principles due to their general character.[35] In addition to ISIS, PYD/YPG has become a significant threat to the security of Türkiye. Although PYD/YPG is the trusted coalition partner of the USA in the fight against ISIS, the direct link between PKK and PYD/YPG has been the biggest problem for USA–Türkiye relations. In 2013, many PKK terrorists from northern Iraq went to Syria to join the PYD/YPG. The goal of the PYD/YPG has been to establish an independent region along the Turkish border.[36]

Turkish Armed Forces (TAF) have gained valuable experience in terms of both technology and doctrine during the counterterrorism activities TAF have carried out since 1984. Although national threat assessments aim to increase the conventional capabilities of the Turkish Army through the influence of international organizations such as NATO and bilateral relations, this irregular warfare situation has prepared TAF for combating hybrid threats, which are considered as the dominant aspect of the contemporary warfare approach. Utilizing valuable experiences acquired in the fight against terrorism in cross-border operations and incorporating them into the organizational culture is a critical example of repurposing existing capability. The fact that these challenging experiences have the potential to permanently alter the armed forces' organizational structure as a learning organization is consistent with the concept of innovation.

Due to the changing security environment, TAF has adopted a flexible structure with high mobility, as has occurred in similar examples throughout the world. Instead of a rigid conventional military structure, TAF tries to keep up with the changes by increasing the number of commando battalions and special forces units that are fit for irregular warfare. TAF additionally carries out training and advisory activities in Iraq, Syria, Azerbaijan, and Libya with the help of the experiences gained in Afghanistan.[37]

The most important contribution Türkiye can make to studies on irregular warfare in the context of military innovation is the experience of anti-terrorist operations against the PKK terrorist organization, which the literature has not covered much. The PKK terrorist organization began its activities in the 1970s. Its first steps included illegal activities such as jewelry robberies and drug trafficking to fund the group. They also engaged in acts of violence against landlords within the sociological and political structure of eastern Türkiye, which they saw as a feudal system.[38] Their objective at that time was to carry out a communist coup based on guerrilla warfare tactics in order to establish an independent Kurdish state. PKK moved toward the Syrian Bekaa Valley with the 1980 military coup and started performing terrorist acts, the victims of which included security forces with the Eruh and Şemdinli raids in August 1984.[39] The most obvious reason for the organization's rapid gain of power and effectiveness is that the government of the period ignored the organization's exploitation of social, economic, and political problems and underestimated its potential and capabilities. In this initial period, the government and military authorities were unable to comprehend the insurgency of the PKK terrorist organization.[40] From this perspective, if the problem could have been handled comprehensively and if an effective intervention had been made, the PKK movement would have ended before it grew.[41] However, the need for coordinated planning and implementation in the economic, cultural, social, legal, political, educational, and administrative fields not just in terms of security forces but also with all state institutions was not understood until the Gulf War.[42]

The general characteristics of the 1980s fight against PKK were identical to those of the suppression operations conducted during the early years of the Turkish Republic. Thus, the pursuit of bandits and terror operations were continued by following a conventional approach and structure.[43] While gendarmerie stations, on one hand, had assumed greater significance in the broader campaign against PKK by the late 1980s. On the other hand, the conventional approach had been abandoned, and small unit operations and the village guard system have been initiated. The field units implemented new tactics, weapons, and structures to counter PKK terrorists effectively. Meanwhile, PKK, which generally used hit-and-run tactics up until the 1990s, became more robust in logistics, weapons, ammunition, and human resources during and after the Gulf War, operating comfortably and safely within northern Iraq. At this time, the armed branch of the PKK (i.e., ARGK) had gained such a capacity that it could stay for several hours in villages and some districts in eastern and southeast Türkiye in order to make armed propaganda.[44] In this way, ARGK reached the capacity to carry out activities such as military outpost raids and road control on critical routes with large groups. Due to the risk of PKK attacks, travel for military personnel and equipment was restricted and maintained mainly by helicopter.[45] For example, the border posts and gendarmerie stations of Derecik, Alan, and Aktütün were attacked by the terrorist organization in 1992, leaving 74 casualties (KIA).[46]

After 1987, when martial law (*Sıkı Yönetim*) that had been ongoing since 1978 transformed into a state of emergency (*Olağanüstü Hal – OHAL*), the PKK terrorist organization increased its violence, and state authorities began to view PKK as a problem that could not be overcome with classical methods. As a result, TAF and the Gendarmerie General Command made significant changes to their weapons, equipment, and structure. This is an important example in terms of adapting irregular warfare to current conditions using an innovative perspective. At this time, the responsibility of the fight against terrorism primarily belonged to the Ministry of the Interior and the Gendarmerie Public Security Command (*Jandarma Asayiş Komutanlığı*). This command was in Diyarbakır and was charged with commanding and coordinating military activities against the PKK terrorist organization in the region under the OHAL. As of 1998, the command was reorganized and only remained responsible for Siirt, Hakkari, Van, and Şırnak provinces. The command was moved to Van and renamed the Gendarmerie Public Security Corps Command *(Jandarma Asayiş Kolordu Komutanlığı)*. TAF was only obliged to provide support upon request; however, the commanders of most gendarmerie units, including the General Commander and Public Security Commander, had been assigned from among TAF's Turkish Land Force Officers.

However, due to the increase in PKK violence, TAF also began to play an active role in the anti-terrorist operations, with some gendarmerie units also coming under the operational command of Land Forces units.[47] As a result of evaluations on how to combat PKK, the Gendarmerie General Command started to increase the number of commando units, beginning at the police station and district level. In the critical regions that needed to be kept under control, fortified gendarmerie stations with a total of 150 people were established, and the region was controlled with patrols and commando posts. When the village guards that had been initiated as a practical project including local people were added to these numbers, powerful structures emerged. The number of commandos in the Gendarmerie had been 7,000 in 1987 and increased to 15,000 in the 1990s. The number of commando battalions increased from ten to 16, the number of commando companies from eight to 84, and the number of gendarmerie stations from 352 to 555.[48] Since most gendarmerie stations had been established to prevent border violations, smuggling, and security incidents, their locations had a serious disadvantage in terms of being able to maintain their own security against PKK raids. This situation necessitated the relocation of a significant part of the gendarmerie stations, or at least additional measures for their security.[49]

The commando units of the Turkish Land Forces were also increased in terms of numbers and structures.[50] Within the Gendarmerie, special units such as A and B teams consisting of professional soldiers instead of conscripts were established, and their numbers increased rapidly.[51] In 1993, the Special Warfare Department was renamed the Special Forces Command and reorganized to carry out irregular warfare activities. The way these smaller, more qualified units were used had also changed compared to the past. First, the small units carrying out operations and patrols in the field attempted to spot terrorist elements and establish contact with them. Following contact, special operations or commando units would reinforce their position by moving into the region and neutralizing the terrorists. These company-level commando units had fire support teams armed with weapons such as 60 mm. mortars, M2 anti-aircraft machine guns, and M18 recoilless rifles.[52] Gendarmerie Internal Security (JIG) Teams, units similar to commando teams, were added to the structure of gendarmerie stations and increased personnel numbers. The stations were also reinforced with 81mm mortars and tanks. Moreover, the number of armored vehicles, attack helicopters, and utility helicopters were increased, as well as their characteristics. Night vision goggles and laser pointers began to be used actively.[53] UH-60 Skorsky helicopters, Russian-made MI-17 and French-made AS-532 Cougar utility helicopters, and AH-1W Super Cobra and 27 AH-1 Cobra attack helicopters were also added to the inventory.[54] Russian-made PK machine guns and SDV Dragunov sniper rifles were also acquired.[55] In addition, the weapons, equipment, and clothes of the military troops were also improved by increasing the budget for the fight against terrorism.[56]

Another important development in the 1990s was the use of cross-border operations and air force elements.[57] In 1992, a joint cross-border operation was launched within Iraq with the participation of 15,000 troops, TAF's second biggest cross-border operation since the Cyprus Peace Operation.[58] As a result of the operation, the PKK terrorist organization had to declare a short-term unilateral ceasefire on March 20, 1993.[59] This first cross-border operation was the signal of TAF's change in perception. Instead of traditional pursuit operations, the military and bureaucratic mechanisms began to evaluate PKK as a problem that must be dealt with using joint operations, including Air Force and Land Force units. The more sophisticated operational plans targeted the terrorist-supporting structures such as camp areas, main routes, and terrorist groups. TAF's cross-border operations have been the most effective method against PKK, and as such have continued to increase.[60]

The so-called unilateral ceasefire was declared a tactic of deception upon being broken on May 24, 1993 with the slaughter of 33 demobilized soldiers traveling unarmed on the Bingöl–Elazığ Highway.[61] On July 5, 1993, another

33 civilians were shot and killed by the terrorists in the village of Başbağlar, Erzincan. Violence continued at the same rate as PKK survived in cross-border camps and safe areas, experiencing no hardships in terms of human resources and outside support. The security forces also increased their capacities and capabilities and continued to increase their effectiveness in the region.

The effectiveness of these operations, along with the capture of so-called PKK leader, Abdullah Öcalan, on July 15, 1999 resulted in a brief period of non-conflict in the fight against terrorism.[62] The number of those who left the PKK between 2003 and 2005 reached 1,500.[63] This positive atmosphere lasted until the 2003 US invasion of Iraq and the overthrow of Saddam Hussein's Ba'ath regime. The safe haven that was recreated for the PKK in Iraq was left uncontrolled and provided an opportunity for them to recover and launch attacks at the outposts across the border.[64] During this period, another ultra-violent faction known as the Kurdistan Freedom Falcons (TAK) emerged. TAK aimed to connect with the youths who found PKK to be too soft for the movement.[65]

After 2005, the terrorist organization tended to increase its effectiveness, this time with handmade explosives (IEDs) and vehicle-borne improvised explosive device (VBIED) attacks. In addition to the internal operations against terrorist groups, as well as the opportunities and capabilities developed in the late 1990s, air operations were carried out against PKK targets in northern Iraq. 2011 may be considered the year when the United States of America directly aided Türkiye in its fight against PKK. American UAVs, the Predators, were requested to hit targets in northern Iraq. The USA approved instant intelligence sharing regarding PKK and the sale of three AH-1 Super Cobra helicopters. Moreover, cooperation was established with Iran through PKK-related intelligence sharing.[66]

Despite not creating a significant change in this new period compared to the innovations experienced in the 1990s, critical technological developments did also take place. For example, replacing BTR and Shortland-type vehicles, which retained protection against light infantry weapons but lacked bottom armor, with the more functional and lighter Otokar COBRA-type vehicles equipped with thermal cameras and weapons that could be controlled from within the vehicle were critical as a safeguard against IED threats.[67] Moreover, more modern vehicle-mounted and backpack jammer technologies were developed as a countermeasure in place of classical frequency mixers. Thermal cameras and thermal weapon binoculars provided TAF with a serious advantage over the terrorists. This technology, the first examples of which were encountered with the *Kaşif* (Explorer) and *Baykuş* (Owl) thermal cameras in the 1990s, started to be used in outposts and gendarmerie stations with the support of the domestic industry between 2006 and 2008. The *Şahin Gözü* (Eagle Eye) thermal camera was the product of this domestic effort. As another positive development, TSD-9450 thermal weapon scopes, the first examples of which were acquired through technology transfer in 1997, were developed by ASELSAN, and *Piton/Boa* thermal weapon scopes with higher resolution capacity were produced in 2007. In 2014, the electronic parts for the MINI-TSD thermal weapon scopes were also updated and built with domestic capabilities. Again, night vision systems that had been acquired through technology transfer in the 1990s were updated as of 2010 into the A300 series.[68]

Despite all these technological advances being integrated into TAF's capabilities, a definite break in the fight against the PKK terrorist organization had yet to be achieved. As of 2011, a solution began to be sought in the fight against terrorism through democratic means, and until 2015, no effective change had been realized.[69] Exploiting this democratic solution process, the PKK terrorist organization increased its activities again in 2015 and this time turned to urban conflicts by creating trenches and barricades in settlements such as Sur in Diyarbakır, Cizre in Şırnak, and Nusaybin in Mardin. This approach was a new PKK tactic that had never been used in a conflict against state forces in an urban area with the presence of civilians. Likewise, urban warfare was a new experience for the security forces. Urban operations, as the most severe and harsh conflicts in military literature, require detailed planning, experience, and training as well as special weapons and equipment. The security forces initially faced challenges as they did not have these capabilities. However, this new era enabled the authorities to be more decisive in the fight against terrorism and solve this problem using all at its disposal with a comprehensive approach, including political, economic, and social aspects.

With increased determination in the fight against terrorism, a high level of coherence has emerged among the Undersecretariat for Defense Industries, domestic manufacturers, and the military units in the field. The R&D departments of industry organizations such as ASELSAN, NUROL, and BAYKAR went directly to the field to determine the needs and demands of the end-users as well as the requirements of the conflict zone on site. As a result of this approach, a very progressive change occurred in the defense industry. The national infantry rifles MPT-76, MPT-55, Bora-12 and KNT-76 sniper rifles, and PMT-76 and SAR-762 MT machine guns replacing the weapons security forces had been using (such as the G3, HK33, and MG3) was an essential part of this change. Mine-resistant wheeled armored vehicle technology, which started with the Cobra vehicle in 1997, was also developed with new models such as the Cobra-II, Kirpi, Kirpi-II, and Ejder-Yalçın. Casualties due to IEDs were minimized in this way. In combating IEDs, search and identification that in the past had only been made with mine detectors and visual searches were replaced by elements equipped with high-tech products such as the Route Mine/IED Clearance (GÜMET), Mine/IED Detection and Disposal (METI), and Explosive Material Reconnaissance and Disposal Team (PMKİ). The need for attack helicopters, which are an essential tool in the fight against terrorism, was also solved by adding T129 ATAK helicopters to the inventory. These helicopters were modified in line with the needs of the Turkish security forces and produced in Türkiye. Also known as the world's most advanced heavy-duty helicopter, the CH-47 Chinook was included in TAF's inventory in 2016, with the number having grown to 11.[70]

Although the security forces suffered severe casualties during the urban warfare operations, they adapted quickly, improved their capabilities, and demonstrated success. After 2015, significant developments were experienced in both the Gendarmerie General Command and TAF. The Gendarmerie General Command, fully under the Ministry of Interior as of 2016, has assumed responsibility for fighting terrorism within Türkiye's borders. With the understanding that the country's security and the fight against terrorism begin beyond the borders of the country, TAF has taken responsibility for cross-border operations. For these purposes, the Gendarmerie General Command quickly increased the number of special operations battalions, commando battalions, and commando companies. The Turkish Land Forces Command increased the number of flexible and highly mobile commando battalions almost tenfold with a similar perspective. As of December 2020, the number of commando brigades within the Turkish Land Forces was around 17. To reach this number, divisions were reduced to brigade level and then transformed into commando brigades.[71] The foundations for this transformation had been lain between 2006 and 2008.[72] Another significant change is that all these units have started being formed from professional soldiers within the body of both the Gendarmerie and the Turkish Land Forces, the plan being in this way to have the transformation occur not only at the quantitative but also at the qualitative level. Although adaptation problems and technical deficiencies have occurred, when considering military institutions' resistance to change, the innovation of security forces can be said to have occurred relatively rapidly.

The most critical factor in this process that has created a serious change in the counterterrorism performance of both the Gendarmerie and TAF is the effective use of uncrewed aerial vehicles (UAVs). The success rate in the battle against terrorism improved substantially once weapon systems were mounted on UAVs, which had first been employed primarily for observation and surveillance. The first UAV to enter TAF inventory was the Banshee system from the British Meggitt company in 1983; it was designed to be used in air force training. Although different drones were used for tactical purposes in the 1990s, the expected effectiveness was unattainable. The GNAT 750s, manufactured by the US General Atomics corporation and initially deployed in 1993, were the first UAVs to be used in counterterrorism operations in Türkiye.

As a result of the USA declining Predator and Raptor UAV sales to Türkiye in 2005, the Israeli-made Heron UAV systems with their long-range and long in-flight time were rented for reconnaissance and surveillance in counterterrorism operations.[73] These new systems put pressure on PKK; however, they also encountered significant problems over time and were unable to meet expectations. As a result of the initiatives of companies such as TUSAŞ, BAYKAR, and VESTEL, UAVs such as TUSAŞ's *Anka, Anka-S,* and *Anka Aksungur*; BAYKAR's *Bayraktar TB-2* and *Akıncı*; and VESTEL's *Karayel* have been developed to address these difficulties and problems using domestic resources. The *Bayraktar TB-2* was armed as of 2015, which achieved a significant advantage. *Anka-S'* ability to be controlled via the TürkSat 4B satellite and carry weapons should also be noted as important developments. *Aksungur* and *Akıncı* UAVs are seen as the systems of the future.

Beyond their technological superiority, the use of these systems regarding coordination with the commandos and special operations units on the ground and their support alongside the Air Force's capabilities reveal innovative development. According to Fukuyama, the prominence of the innovation UAVs has created in this way can only be compared with the invention of HMS Dreadnought in naval history and aircraft carriers during WWII. Moreover, due to UAVs being economical and difficult to destroy as well as preventing human casualties, they have changed the character of warfare against complex and expensive weapon systems such as tanks, helicopters, and aircraft.[74] The capabilities of these UAVs are also supported by the development of the *Hürkuş* light attack aircraft. Being designed as an American A-29 Super Tucano-style aircraft and developed to combat terrorism and insurgency, the *Hürkuş-C's* efficient use against ground targets has made significant impact in the fight against terrorism.[75]

The most important experience TAF gained during the ongoing irregular war in Syria is considered to be the effective use of these UAVs. The Anka-S and TB2 Bayraktar UAVs that are in TAF's inventory have been effectively used within the scope of internal security operations. The deaths of 33 soldiers due to the Assad regime's attack, backed by the Russian Federation, on Turkish checkpoints and the convoy located in the region between Balioun and Al-Barah villages south of Idlib on February 22, 2020, caused Türkiye to retaliate against the Syrian administration.[76] During this retaliation, five helicopters, 23 tanks, 23 artillery batteries, and Russian-made Buk and Pantsir air defense systems were destroyed. This was the first case in which the international actors recognized the effectiveness of Turkish UAVs. Türkiye, which lost only four UAVs in this process, has become one of the few countries globally to have domestic UAV production and to have used it in practice. Turkish UAVs were also successfully used to support the UN-recognized Libyan Government of National Accord against the opposition leader Khalifa Hafter's Libyan National Army. Turkish UAVs have had immense success in the victory of the Azerbaijani army against Armenia during the Nagorno-Karabakh War. The UAV attacks that destroyed many Armenian targets (i.e., 200 tanks, 90 armored cars, and 182 artillery units) guaranteed Türkiye's elevated position in the region.[77] UAVs, Turkish Koral electronic warfare capabilities, and Korkut short-range air defense systems were also used in Libya. These systems were highly effective in countering the HAG's aviation assets maintained by private security companies and their sponsoring states.

TAF's development of irregular warfare capabilities has also been used effectively as a tool of foreign policy. For example, training and advising missions for the forces affiliated with the Libyan Government of National Accord (e.g., infantry units, coast guard, special forces, forward observation officers, and artillery systems) were carried out in Libya.[78] Somalia is another country Türkiye has supported with security sector reforms and security force assistance.[79]

With the use of TAF units across borders, having easily transportable secure base areas that are also able to be built in a short time came to the agenda. The security tower systems with living areas used to combat terrorism also need to be mentioned. These tower systems have a diameter of six meters and a height of 21 meters; they are deployed at critical points and offer uninterrupted surveillance.[80] Another critical project is the modular temporary base system developed by ASELSAN. These bases can be established quickly and aim to ensure the dominance of security forces in critical areas. The system is equipped with living units reinforced against mortar fire; they have ballistic protection, surveillance, remote-controlled weapon, shooting location detection, early warning, and electronic warfare systems as well as land surveillance radar and UAV systems.[81] Commando units supported by UAVs and air elements are also endowed with tanks in cross-border operations. For example, M60 tanks were modernized by Israel with the Sabra MK II package and renovated in terms of weapons, electronic systems, and engines. After this renovation, the M60Ts were once again modernized by ASELSAN with the *Pulat* active armor technology and other electronic systems in 2017 and turned into the M60TM. The experience gained from cross-border and urban warfare operations have contributed significantly to these modernizations.[82]

Innovation does not just refer to development in military technology. Strategic and doctrinal change should accompany this technological development. This situation is observable in the example of the Turkish fight against terrorism. State authorities and soldiers evaluated the PKK terrorist organization as a security issue in the beginning. The precautions and strategy were shaped around this perception. In the late 1990s, PKK was understood to be exploiting the social, economic, and political problems of the region; as such, the government needed a comprehensive approach that involved all institutions and capabilities in the struggle against PKK terrorism. This new perception has certain similarities to the hearts and minds approach. The same mindset has been implemented in northern Iraq and northern Syria in contemporary cross-border operations. The end goal of the Turkish interventions in both countries is to create a protected and secure zone safe from terrorist organizations along the border area. For this reason, Türkiye has not only used its military presence in this region but also its law enforcement agencies to build law and order and civilian assets to foster economic and political development.[83] The Turkish Gendarmerie and the Turkish police have conducted security sector reform activities to build local law enforcement agencies. These activities can be determined as part of Türkiye's understanding of irregular warfare against terrorist organizations such as PKK/YPG and ISIS in the region. As mentioned earlier, stability policing is a rising trend in the NATO doctrine, and Türkiye's efforts can be seen as an example of this trend.

TAF experienced a serious change in doctrine and technology in becoming a regional power. The domestically produced UAV technology has brought an advantage against terrorist and insurgency groups as well as in international conflicts. Apart from this vital capability, building the necessary military structure and coordination and determining appropriate political goals are also important aspects. Administering this change should be shaped by scientific evaluations and analytical projections of the future. Therefore, ensuring the sustainability of studies carried out by research centers able to evaluate the experiences gained in such wars, conflicts, and military technology endeavors is important from a critical point of view. This critical thinking enables decision-making mechanisms to predict what kind of preparations can be conducted in the future. Innovation, it should be remembered, is not just about introducing new technology but also involves issues such as how new technology can be used in the best way and how to integrate it with existing capabilities.

Conclusion

Emphasizing innovation as a word that does not just refer to technological advancement while also emphasizing new technology to be a factor that can and should promote changes in strategy and organizational structure is imperative. Therefore, the focus should not be on the short-term effects of emerging weapon systems and military technologies. As noted by General George S. Patton, Jr., "Wars may be fought with weapons, but they are won by men".[84] Although the dominance of new technologies seems inevitable in wars and conflicts, the evaluation has been made that human element will continue to maintain its importance in the planning and implementation stages.

The concept of military innovation needs an organizational culture that is open to creative thinking and change. This approach breaks down existing thought patterns beyond testing or proves that a conceived solution will work. At the same time, it includes the reinterpretation of current possibilities and capabilities through different perspectives and by discovering new technologies. Innovation is seen as a continuous process and necessitates an enduring evaluation of changing threat perceptions, past experiences, developments in other examples, and the characteristics of the tasks to be encountered.

As seen in historical examples, IW is not only a concept that enables the weak side to fight against the strong. With the changing character of war, IW has become an approach that should be adopted by conventional armies that have to encounter armed non-state groups and terrorist and extremist organizations, as well as the conventional capacities of potential adversary states. Although the methodologies and tactics of IW seem like a primitive form of warfare, it still preserves vitality and certainty on the battleground. For instance, despite the developing weapon systems and technological improvements in defense industries, AK-47s and IEDs are still the most effective elements in the hands of a decisive and capable adversary for fighting against the strongest armies, as seen in the case of the Taliban insurgency in Afghanistan. Therefore, IW will maintain its validity and effectiveness in future wars. From this perspective, IW will continue to be perceived as an innovative and creative idea in the future war environment that necessitates a high-level of flexibility and adaptation capability.

Notes

1 Jeremy Black, *Introduction to Global Military History: 1775 to the Present Day*, Routledge, London, 2018, pp. 72–73.
2 Theo Farrell and Terry Terriff, "The Sources of Military Change", in (eds.) Theo Farrell and Terry Terriff *The Sources of Military Change: Culture, Politics, Technology*, Lynne Rienner Publishers, London, 2002, pp. 10–11.
3 Lawrence Freedman, *Strategy: A History*, Oxford University Press, 2013, p. 10.
4 *JP 1, Doctrine for the Armed Forces of the United States*, Joint Chiefs of Staff, Washington, 2007, p. 27.
5 Eric V. Larson et al., *Assessing Irregular Warfare: A Framework for Intelligence Analysis*, RAND, Santa Monica, 2008, pp. 8–9.
6 *JP 3-05, Special Operations*, Joint Chiefs of Staff, Washington, 2014.
7 Otto Heilbrunn, *Warfare in the Enemy's Rear*, Routledge, London, 1963, p. 33.
8 Max Boot, *Invisible Armies: An Epic History of Guerrilla Warfare from Ancient Times to the Present*, 1st ed., Liveright Pub. Corp., New York, 2013, pp. 258–259.
9 John A. Nagl, "Counterinsurgency in Vietnam: American Organizational Culture and Learning", in (eds.) D. Marston and C. Malkasian, *Counterinsurgency in Modern Warfare*, Osprey Publishing, Oxford, 2008, p. 132.
10 Austin Long, *Doctrine of Eternal Recurrence the U.S. Military and Counterinsurgency Doctrine, 1960–1970 and 2003–2006*, RAND, Santa Monica, 2008, pp. 12–13.
11 David Galula, *Counterinsurgency Warfare; Theory and Practice*, Praeger, New York, 1964, p. 65.
12 Dale A. Andradé and James H. Willbanks, "CORDS/Phoenix: Counterinsurgency Lessons from Vietnam for the Future", *Military Review*, 86, 2006, p. 14.
13 Thomas W. Scoville, *Reorganizng for Pacification Support*, Center of Military History United States Army, Washington, 1999.
14 Black, *Introduction to Global Military History: 1775 to the Present Day*, p. 214.
15 James H. Hay, *Tactical and Material Innovations*, Department of the Army, Washington, 2002, p. 9.
16 Hay, *Tactical and Material Innovations*, pp. 19–20.
17 Antonio Giustozzi, *Koran, Kalashnikov and Laptop: The Neo-Taliban Insurgency in Afghanistan*, Hurst & Company, London, 2007.
18 James A. Russell, "Counterinsurgency American Style: Considering David Petraeus and Twenty-First Century Irregular War", *Small Wars & Insurgencies*, 25:1, 2014.
19 Theo Farrell, "Improving in War: Military Adaptation and the British in Helmand Province, Afghanistan, 2006–2009", *Journal of Strategic Studies*, 33:4, 2010, pp. 567–94.
20 US Department of Defense, *Quadrennial Defense Report*, ed. Department of Defense, Washington, 2010, p. 10.
21 Farrell, "Improving in War: Military Adaptation and the British in Helmand Province, Afghanistan, 2006–2009".

22 Theo Farrell, Sten Rynning, and Terry Terriff, *Transforming Military Power since the Cold War: Britain, France, and the United States, 1991–2012*, Cambridge University Press, 2013, pp. 60–85.
23 Marshall Curt Erwin, *Intelligence, Surveillance, and Reconnaissance (ISR) Acquisition: Issues for Congress, Report*, Congressional Research Service, Washington, 2013.
24 Rita Boland, "Air Force ISR Changes After Afghanistan", *Signal Magazine*, 2014, https://bit.ly/2U9comV, accessed on 01.05.2014.
25 Robert P. Haffa and Anand Datla, "Joint Intelligence, Surveillance, and Reconnaissance in Contested Airspace", *Air and Space Power Journal*, 28:3, 2014, p. 31.
26 Michael C. Horowitz, Joshua A. Schwartz, and Matthew Fuhrman, "China Has Made Drone Warfare Global", *Foreign Affairs*, 2020, https://fam.ag/3vYuVjd, accessed on 20.10.2020.
27 Sharon Weinberger, "China Has Already Won the Drone Wars", *Foreign Policy*, 2021, https://bit.ly/3x1nHMM, accessed on 10.05.2018.
28 Horowitz, Schwartz, and Fuhrman, "China Has Made Drone Warfare Global".
29 Andrew Self and Jared Ferris, "Dead Men Tell No Lies: Using Killed-in-Action (KIA) Data to Expose the PKK's Regional Shell Game", *Defence Against Terrorism Review DATR*, 8:1, 2016, p. 25.
30 Alexander Rabin, "Diplomacy and Dividends: Who Really Controls the Wagner Group?", *Foreign Policy Research Institute, Intern Corner*, 2019), https://bit.ly/3h98oL5, accessed on 04.10.2019.
31 Amy Mackinnon, "Russia's Shadowy Mercenaries Offer Humanitarian Aid to Clean Image", *Foreign Policy*, 2021, https://bit.ly/2U4Xe2m, accessed on 22.07.2020.
32 UN Panel of Experts Established Pursuant to Security Council Resolution 1973, "Letter Dated 8 March 2021 from the Panel of Experts on Libya Established Pursuant to Resolution 1973 (2011) Addressed to the President of the Security Council", 2021, p. 32.
33 UN Panel of Experts Established Pursuant to Security Council Resolution 1973, pp. 30–31.
34 Giuseppe De Magistris and Stefano Bergonzini, "Stability Policing: A Golden Opportunity for NATO", *Romanian Gendarmerie*, June 2020.
35 Michael Knights and Wladimir van Wilgenburg, *Accidental Allies*, I.B. Tauris, London, 2021, p. 3.
36 Knights and van Wilgenburg, *Accidental Allies*, p. 29.
37 Knights and van Wilgenburg, *Accidental Allies*, pp. 9–10.
38 Paul White, *The PKK: Coming Down from the Mountains*, Rebels, Zed Book, London, 2015, p. 102.
39 Nur Bilge Criss, "The nature of PKK terrorism in Turkey", *Studies in Conflict & Terrorism*, 18:1, 1995, p. 18.
40 Ahmet Özcan, *Operasyon Var Bu Gece: Bir Komando Asteğmenin Güneydoğu Hatıraları*, Kent Kitap, Ankara, 2008, p. 57.
41 İsmet G. İmset, *The PKK: A Report on Separatist Violence in Turkey (1973–1992)*, Turkish Daily News Publication, Ankara, 1992, p. 30.
42 Mehmet Ali Kışlalı, *Güneydoğu: Düşük Yoğunluklu Çatışma*, Ümit Yayıncılık, Ankara, 1996, p. 167.
43 Kışlalı, *Güneydoğu: Düşük Yoğunluklu Çatışma*, p. 179.
44 White, *The PKK: Coming Down from the Mountains*, p. 106.
45 Özcan, *Operasyon Var Bu Gece: Bir Komando Asteğmenin Güneydoğu Hatıraları*, p. 61.
46 Aziz Ergen, "Terörle Mücadelede Eşref Bitlis'ten Alınan Dersler", (2019), https://bit.ly/3A41hwo, accessed on 02.02.2019.

47 Naci Akkaş, "Jandarma Teşkilatına İlişkinin Genel Bilgiler", in (ed.) Aydın Ziya Özgür, *Jandarma Görev ve Yetkileri*, Anadolu Üniversitesi Yayınları, Eskişehir, 2004.
48 Eşref Bitlis, "Güneydoğu'da 25 Bin Komando İşbaşında", interview by Mehmet Ali Kışlalı, 1996, pp. 230–239.
49 Osman Pamukoğlu, *Unutulanlar Dışında Yeni Bir Şey Yok*, İnkılap Kitabevi, Ankara, 2010.
50 Pamukoğlu, *Unutulanlar Dışında Yeni Bir Şey Yok*.
51 Bitlis, "Güneydoğu'da 25 Bin Komando İşbaşında" pp. 230–239.
52 Özcan, *Operasyon Var Bu Gece: Bir Komando Asteğmenin Güneydoğu Hatıraları*, p. 87.
53 Bitlis, "Güneydoğu'da 25 Bin Komando İşbaşında", pp. 230–239.
54 Bitlis, "Güneydoğu'da 25 Bin Komando İşbaşında"; Ergen, "Terörle Mücadelede Eşref Bitlis'ten Alınan Dersler".
55 Güreş, "Terörü Kısa Sürede Sona Erdireceğiz", p. 222.
56 Özcan, *Operasyon Var Bu Gece: Bir Komando Asteğmenin Güneydoğu Hatıraları*, p. 89.
57 Sabri Yirmibeşoğlu, "Tedbir Geç Alındı", interview by Mehmet Ali Kışlalı, *Güneydoğu: Düşük Yoğunluklu Çatışma*, 1996, p. 242.
58 Can Kasapoğlu, "Assessing the Role Of Cross-Border Military Operations In Confronting Transnational Violent Non-State Groups: 1992–1998 Turkish Armed Forces Case", *Defence Against Terrorism Review DATR*, 4:1, 2012, p. 76.
59 Ergen, "Terörle Mücadelede Eşref Bitlis'ten Alınan Dersler".
60 Kasapoğlu, "Assessing the Role Of Cross-Border Military Operations In Confronting Transnational Violent Non-State Groups: 1992–1998 Turkish Armed Forces Case", p. 78.
61 Erdal Sarızeybek, *İhaneti Gördüm*, Pozitif Yayınları, İstanbul, 2007.
62 Mehmet Gurses, *Anatomy of a Civil War*, University of Michigan Press, Ann Arbor, 2018, p. 6.
63 White, *The PKK: Coming Down from the Mountains*, p. 76.
64 Gurses, *Anatomy of a Civil War*, p. 6.
65 White, *The PKK: Coming Down from the Mountains*, p. 130.
66 White, *The PKK: Coming Down from the Mountains*, pp. 194–95.
67 Ahmet Alemdar, "Türk Zırhlısı Cobra Savaşta: Güney Osetya Savaşı", *Defence Türk*, 2019, https://bit.ly/3xSwyQO, accessed on 29.05.2019.
68 Emin Y. Köksal and Neşe K. Yazar, "Silah Üstü Elektro-Optik Sistemler", *Aselsan Dergi*, 107, 2020, pp. 78–81.
69 Gurses, *Anatomy of a Civil War*, p. 7.
70 "Türkiye, 4 Adet Boeing CH-47F Chinook Helikopteri Daha Teslim Aldı", *Independent Türkçe*, 2019, https://bit.ly/3wTfD0h, accessed on 31.07.2019.
71 Ali Kemal Erdem, "TSK, Komandoya Ağırlık Verdi, Bu Alandaki Gücünü 10 Kat Artırdı", *Independent Türkçe*, 2020, https://bit.ly/3h4TG7S, accessed on 10.12.2020.
72 İlker Başbuğ, *Suçlamalara Karşı Gerçekler*, Kaynak Yayınları, İstanbul, 2014.
73 Francis Fukuyama, "Droning On in the Middle East", *American Purpose*, 2021, https://bit.ly/35Nkq7A, accessed on 05.04.2021.
74 Fukuyama, "Droning On in the Middle East".
75 "Hürkuş-C", Turkish Aerospace, 2021, https://bit.ly/3jahRV6, accessed on 27.05.2021.
76 Cihat Arpacık and Abdurrahman Koç, "İdlib Saldırısı", *Independent Türkçe*, 2020, https://bit.ly/2TUzdLb, accessed on 28.02.2020.
77 Fukuyama, "Droning On in the Middle East".
78 "Letter Dated 8 March 2021 from the Panel of Experts on Libya Established Pursuant to Resolution 1973 (2011) Addressed to the President of the Security Council", pp. 22–23.

79 Burak Dağ, "Turkey Hands Over Military Barracks to Somalia", *Anadolu Ajansı*, 2021, https://bit.ly/3vzyRAq, accessed on 06.03.2021.
80 Yunus Okur and Haydar Toprakçı, "Terörle Mücadelede Yeni Konsept: Yaşam Alanlı Güvenlik Kuleleri", *Anadolu Ajansı*, 2019, https://bit.ly/3gTclEP, accessed on 22.03.2019.
81 Göksel Yıldırım, "TSK'ya 4 Yeni Modüler Üs Bölgesi", *Anadolu Ajansı*, 2019, https://bit.ly/3daPkey, accessed on 02.02.2019.
82 "M60T Tank Modernizasyonları Tamamlandı, M60T Yeni Teknoloji ve Kabiliyetler Kazandı", Savunma Sanayii Başkanlığı, 2020, https://bit.ly/2UEMmbP, accessed on 12.07.2020.
83 Büşra Nur Yılmaz, Sinan Balcıkoca, and Cem Şan, "İçişleri Bakanı Soylu: 21. Yüzyılın Gelişmiş Medeniyetleri Orta Doğu'da Teröre Senaristlik ve Rejisörlük Yapmaktadır", *Anadolu Ajansı*, 2020, https://bit.ly/35TawBJ, accessed on 15.10.2020.
84 George S. Patton Jr., "Mechanized Forces," *the Cavalry Journal*, 42:79, September–October 1933, p. 8.

References

Alemdar, Ahmet, "Türk Zırhlısı Cobra Savaşta: Güney Osetya Savaşı", *Defence Türk*, 2019, https://bit.ly/3xSwyQO.
Akkaş, Naci, "Jandarma Teşkilatına İlişkinin Genel Bilgiler", in (ed.) Aydın Ziya Özgür, *Jandarma Görev ve Yetkileri*, Anadolu Üniversitesi Yayınları, Eskişehir, 2004.
Andradé, Dale A. and Willbanks, James H., "CORDS/Phoenix: Counterinsurgency Lessons from Vietnam for the Future", *Military Review*, 86, 2006, pp. 9–23.
Arpacık, Cihat and Koç, Abdurrahman, "İdlib Saldırısı", *Independent Türkçe*, 2020, https://bit.ly/2TUzdLb, accessed on 28.02.2020.
Başbuğ, İlker, *Suçlamalara Karşı Gerçekler*, Kaynak Yayınları, İstanbul, 2014.
Black, Jeremy, *Introduction to Global Military History: 1775 to the Present Day*, Routledge, London, 2018.
Bitlis, Eşref, "Güneydoğu'da 25 Bin Komando İşbaşında", interview by Mehmet Ali Kışlalı, 1996.
Boland, Rita, "Air Force ISR Changes After Afghanistan", *Signal Magazine*, 2014, https://bit.ly/2U9comV.
Boot, Max, *Invisible Armies: An Epic History of Guerrilla Warfare from Ancient Times to the Present*, 1st ed., New York, Liveright Pub. Corp., 2013.
Criss, Nur Bilge, "The Nature of PKK Terrorism in Turkey", *Studies in Conflict & Terrorism*, 18:1, 1995, pp. 17–37.
Dağ, Burak, "Turkey Hands Over Military Barracks to Somalia", *Anadolu Ajansı*, 2021, https://bit.ly/3vzyRAq.
De Magistris, Giuseppe and Bergonzini, Stefano, "Stability Policing: A Golden Opportunity for NATO", *Romanian Gendarmerie*, June 2020.
Erdem, Ali Kemal, "TSK, Komandoya Ağırlık Verdi, Bu Alandaki Gücünü 10 Kat Artırdı", *Independent Türkçe*, 2020, https://bit.ly/3h4TG7S.
Ergen, Aziz, "Terörle Mücadelede Eşref Bitlis'ten Alınan Dersler," (2019), https://bit.ly/3A41hwo.
Erwin, Marshall Curt, *Intelligence, Surveillance, and Reconnaissance (ISR) Acquisition: Issues for Congress, Report*, Congressional Research Service, Washington, DC, 2013.
Farrell, Theo, "Improving in War: Military Adaptation and the British in Helmand Province, Afghanistan, 2006–2009", *Journal of Strategic Studies*, 33:4, 2010, pp. 567–594.

Farrell, Theo, Rynning, Sten, and Terriff, Terry, *Transforming Military Power since the Cold War: Britain, France, and the United States, 1991–2012*, Cambridge University Press, Cambridge, 2013.

Farrell, Theo and Terriff, Terry "The Sources of Military Change", in (eds.) Theo Farrell and Terry Terriff *The Sources of Military Change: Culture, Politics, Technology*, Lynne Rienner Publishers, London, 2002, pp. 3–20.

Freedman, Lawrence, *Strategy: A History*, Oxford University Press, New York, 2013.

Fukuyama, Francis, "Droning On in the Middle East," *American Purpose*, 2021, https://bit.ly/35Nkq7A

Galula, David, *Counterinsurgency Warfare; Theory and Practice*, Praeger, New York, 1964.

Giustozzi, Antonio Koran, *Kalashnikov and Laptop: The Neo-Taliban Insurgency in Afghanistan*, Hurst & Company, London, 2007.

Gurses, Mehmet, *Anatomy of a Civil War*, University of Michigan Press, Ann Arbor, 2018.

Haffa, Robert P. and Datla, Anand, "Joint Intelligence, Surveillance, and Reconnaissance in Contested Airspace", *Air and Space Power Journal*, 28:3, 2014, pp. 29–47.

Hay, James H., *Tactical and Material Innovations*, Department of the Army, Washington, DC, 2002.

Heilbrunn, Otto, *Warfare in the Enemy's Rear*, Routledge, London, 1963.

Horowitz, Michael C., Schwartz, Joshua A., and Fuhrman, Matthew, "China Has Made Drone Warfare Global," *Foreign Affairs*, 2020, https://fam.ag/3vYuVjd

"Hürkuş-C", Turkish Aerospace, 2021, https://bit.ly/3jahRV6

İmset, İsmet G., *The PKK: A Report on Separatist Violence in Turkey (1973–1992)*, Turkish Daily News Publication, Ankara, 1992.

JP 1, Doctrine for the Armed Forces of the United States, Joint Chiefs of Staff, Washington, DC, 2007.

JP 3-05, Special Operations, Joint Chiefs of Staff, Washington DC, 2014.

Kasapoğlu, Can, "Assessing The Role Of Cross-Border Military Operations In Confronting Transnational Violent Non-State Groups: 1992–1998 Turkish Armed Forces Case," *Defence Against Terrorism Review DATR*, 4:1, 2012, pp. 71–84.

Kışlalı, Mehmet Ali, *Güneydoğu: Düşük Yoğunluklu Çatışma*, Ümit Yayıncılık, Ankara, 1996.

Knights, Michael and van Wilgenburg, Wladimir, *Accidental Allies*, I. B. Tauris, London, 2021.

Köksal, Emin Yiğit and Yazar, Neşe Kılıç, "Silah Üstü Elektro-Optik Sistemler", *Aselsan Dergi*, 107, 2020.

Larson, Eric V. et al., *Assessing Irregular Warfare: A Framework for Intelligence Analysis*, RAND Corporation, Santa Monica, 2008.

Long, Austin, *Doctrine of Eternal Recurrence the U.S. Military and Counterinsurgency Doctrine, 1960–1970 and 2003–2006*, RAND Corporation, Santa Monica, 2008.

Mackinnon, Amy, "Russia's Shadowy Mercenaries Offer Humanitarian Aid to Clean Image", *Foreign Policy*, 2021, https://bit.ly/2U4Xe2m

"M60T Tank Modernizasyonları Tamamlandı, M60T Yeni Teknoloji ve Kabiliyetler Kazandı", Savunma Sanayii Başkanlığı, 2020, https://bit.ly/2UEMmbP

Nagl, John A., "Counterinsurgency in Vietnam: American Organizational Culture and Learning", in (eds.) D. Marston and C. Malkasian, *Counterinsurgency in Modern Warfare*, Osprey Publishing, Oxford, 2008.

Okur, Yunus and Toprakçı, Haydar, "Terörle Mücadelede Yeni Konsept: Yaşam Alanlı Güvenlik Kuleleri", *Anadolu Ajansı*, 2019, https://bit.ly/3gTclEP

Özcan, Ahmet, *Operasyon Var Bu Gece: Bir Komando Asteğmenin Güneydoğu Hatıraları*, Kent Kitap, Ankara, 2008.

Pamukoğlu, Osman, *Unutulanlar Dışında Yeni Bir Şey Yok*, İnkılap Kitabevi, Ankara, 2010.

Patton, George S. Jr., "Mechanized Forces", *the Cavalry Journal*, 42:79, September-October 1933, pp. 5–8.

Rabin, Alexander, "Diplomacy and Dividends: Who Really Controls the Wagner Group?," *Foreign Policy Research Institute, Intern Corner*, 2019), https://bit.ly/3h98oL5

Russell, James A., "Counterinsurgency American style: Considering David Petraeus and Twenty-First Century Irregular War", *Small Wars & Insurgencies*, 25:1, 2014, pp. 69–90.

Sarızeybek, Erdal, *İhaneti Gördüm*, Pozitif Yayınları, İstanbul, 2007.

Scoville, Thomas W., *Reorganizng for Pacification Support*, Center of Military History United States Army, Washington, DC, 1999.

Self, Andrew and Ferris, Jared, "Dead Men Tell No Lies: Using Killed-in-Action (KIA) Data to Expose the PKK's Regional Shell Game", *Defence Against Terrorism Review DATR*, 8:1, 2016, pp. 9–35.

"Türkiye, 4 Adet Boeing CH-47F Chinook Helikopteri Daha Teslim Aldı", *Independent Türkçe*, 2019, https://bit.ly/3wTfD0h

UN. Panel of Experts Established Pursuant to Security Council Resolution 1973, "Letter Dated 8 March 2021 from the Panel of Experts on Libya Established Pursuant to Resolution 1973 (2011) Addressed to the President of the Security Council", 2021.

U.S. Department of Defense, *Quadrennial Defense Report*, ed. Department of Defense, Department of Defense, Washington, DC, 2010.

Weinberger, Sharon, "China Has Already Won the Drone Wars", *Foreign Policy*, 2021, https://bit.ly/3x1nHMM

White, Paul, *The PKK: Coming Down from the Mountains*, Zed Book, London, 2015.

Yirmibeşoğlu, Sabri, "Tedbir Geç Alındı", interview by Mehmet Ali Kışlalı, *Güneydoğu: Düşük Yoğunluklu Çatışma*, 1996.

Yıldırım, Göksel, "TSK'ya 4 Yeni Modüler Üs Bölgesi," *Anadolu Ajansı*, 2019, https://bit.ly/3daPkey

Yılmaz, Büşra Nur, Balcıkoca, Sinan and Şan, Cem, "İçişleri Bakanı Soylu: 21. Yüzyılın Gelişmiş Medeniyetleri Orta Doğu'da Teröre Senaristlik ve Rejisörlük Yapmaktadır", *Anadolu Ajansı*, 2020, https://bit.ly/35TawBJ

7 The Transformation of the Turkish Armed Forces

Lessons Learned in Counterterrorism Operations

Tolga Ökten

Introduction

Turkish Armed Forces (TAF) were a mass conscripted force that was formed, equipped, and trained against a mass Soviet invasion throughout the Cold War. In the aftermath of the Cold War, TAF's threat perception changed, with its organizational structure, doctrine, and weapon systems all starting at the same time to transform for a new threat. This threat was the terrorist organization PKK/KCK[1], and TAF quickly realized they could not be handled with traditional force structures. This transformation deserves scholarly attention with regard to exploring and understanding the Turkish experience. Accordingly, this chapter aims to explore TAF's transformation during its counterterrorism operations in three parts. First, the theoretical framework used to conceptualize military transformation is explained. The theory is built on two essential concepts that constitute combat capacity: fire and maneuver. Second, TAF's transformation is explained in light of this theoretical framework. Last, the reflections of this transformation onto the counterterrorism strategy are analyzed.

Conceptual Framework – Fire and Maneuver

During the 20th century, military change was studied by British historians, Soviet military personnel, and American researchers under the labels of Military Revolution (MR)[2], Military Technical Revolution (MTR)[3], and the best-known, Revolution in Military Affairs (RMA).[4] Since the new millennium, this change has been called Military Transformation.[5] The champion of this school is Donald Rumsfeld, and according to him, this type of transformation is much more profound than new advanced technological weapon systems. To accomplish Military Transformation, technology as well as the way the military thinks, trains, exercises, and fights need to be changed.[6] The main objective of military transformation is to enhance combat power, which consists of the capacities of maneuverability and firepower.[7] This is why TAF's transformation can be studied under these two essential elements of combat power.

DOI: 10.4324/9781003327127-8

The Fire Dimension

Firepower refers to the technology that is used to hit and destroy/neutralize targets. In addition to its destructive power, other technical variables such as range, accuracy, and fire rate are also critical attributes of firepower.[8] At the tactical level, firepower supports maneuver units, and at the strategic level, it can be used to win wars, at least in theory.

Until the 20th century, firepower was almost entirely provided by land-based weapons.[9] Longbows, siege engines, artillery, and machine guns have always been vital firepower instruments. However, the concept of firepower has dramatically changed since the early phases of World War II. Airpower has become the most important firepower source, at both tactical and strategic levels. Because range limitations have become obsolete, city centers have become the first to get hit rather than safe zones for civilians.

Today, airpower is the principal instrument of firepower and is represented by precision strike capability, which enables targets that are fixed in time and space to be accurately neutralized. Modern combat power depends on this capacity, which is called the find-fix-fight sequence. This sequence is also commonly called a kill chain or sensor-to-shooter sequence and is composed of intelligence-surveillance-reconnaissance (ISR) assets, command-control-communications-computer (C4) infrastructure, and precision-guided munitions (PGMs). ISR comprises technical assets like IMINT and SIGINT, which find and fix targets. The raw intelligence coming from these sources is analyzed, synthesized, and shared with the help of C4 technology in a network-centric model. The last part of this chain has PGMs completing the sequence.[10] Precision engagement has become a standard capability for modern military forces. In fact, PGM use on the battlefield has gradually grown since the Gulf War. For instance, just 6% of the ammunition used in the Gulf War was a PGM; in contrast, this percentage was 29% in Kosovo, 57% in Afghanistan, and 100% in Libya.[11]

Technology enables this chain to work nearly simultaneously.[12] Especially after the introduction of UAV technology, the wall between sensor and shooter disappeared, and intelligence and operation functions were merged.[13] Technology can be argued to be at the center of this model and the driving force behind the military transformation. Precision engagement triggered a transformation in the organizational structures of maneuver units and doctrines.

The Maneuver Dimension

Maneuver involves the movement of military units during or before battles to gain a relatively advantageous position against the enemy.[14] In a conventional maneuver, the goal is to hold ground.[15] Due to the nature of the mission, it must be performed by ground troops and necessitates intensive close-quarter combat. Due to logistical constraints, it is often conducted at the tactical and operational levels.

Fire and maneuver are inextricably linked. As firepower's lethality increases, maneuver tactics transform. Before the late 19th century, regular army tactical engagements were conducted by soldiers with shiny colored uniforms, moving in unconcealable columns with standard infantry engagements such as volley fire and bayonet attacks. The units with small, agile, and stealthy elements positioned at the rear of the tactical or operational levels were called irregulars.[16] This started to change with the increasing lethality of firepower. In the American Civil War, Napoleonic tactics were seen to have become too dangerous against firepower. In 1914, moving the front line was practically impossible, which turned World War I into a war of attrition.

The critical turn for maneuver warfare was the tactical and organizational innovations that were initiated to overcome this trench stalemate. These innovations marked post-1918 battles. According to Murray, a soldier in 1914 would have had no idea about the operational concepts of 1918, but a soldier from 1918 would have understood the tactics of World War II and even the Gulf War.[17]

For example, the British tried combining the infantry and tank arms of frontal attacks. To counter this, Germans tried penetration and encirclement tactics to outflank the enemy's tactical positions instead of full-scale frontal attacks. *Sturmtruppen* [Stormtrooper] units were the main maneuver elements in these tactics, teams of 4 to 12 people with weapons such as hand grenades, flame throwers, and submachine guns instead of the classical rifle-bayonet duo. These units were supported by artillery. To achieve surprise, artillery components were secretly transported to the tactical level prior to the infantry breakthrough.[18] This new doctrine, later known as the "blitzkrieg without tanks",[19] distinguished itself by successfully integrating the firepower and maneuver components and laying the groundwork for contemporary infantry tactics.

Firepower's range and kill box increased dramatically throughout the 20th century. Although this was mainly due to airpower, artillery also progressed. Between 1900 and 1980, the range of artillery increased from 10 kilometers to 40, and the lethal area increased from 1000 to 20000 square meters incrementally.[20] Due to the negative correlation existing between lethality of firepower and troop density, armies avoided mass movements and instead enlarged the battleground radius. From antiquity to contemporary times, the force-to-space ratio of a battleground has decreased, and the length and depth of the front lines has increased continuously. For example, in the Napoleonic Wars the average number of men per square kilometer was 4,790. These numbers were 404 in World War I, 36 in the World War II, and 2.34 in the Gulf War.[21] Accordingly, battles have evolved into a sequence of tactical engagements of small units rather than a decisive one.[22] This phenomenon can be called the *irregularization* of regular armies.

Tactical and operational maneuvers aim to kill enemies while surviving under intense and precise fire; this is referred to as the combination of lethality and survivability.[23] This is valid not only for units at the tactical level but also for logistical and reserve units.[24] According to Biddle, in order to balance lethality and survivability, military units should use the *modern system of force employment*. The basic rules of the modern system are: *exposure-reduction tactics of cover, concealment, dispersion, small-unit independent maneuver, suppression,* and *combined-arms integration* at the tactical level and *breakthrough, exploitation, depth, reserves,* and *counterattack* at the operational level. Biddle claimed these were the common points of thousands of pages of manuals and doctrines.[25]

The most critical question is why do all militaries not practice the *modern system of force employment* at the same level, if it is so important for modern warfare? The answer is that it requires training, experience, expertise, and talented people. Rules like *small-unit independent maneuver* and *dispersion* require the initiative and leadership of low-rank tactical commanders. *Low-exposure tactics* and *concealment* require muscle memory and tactical skills.[26] That is why force employment dictates volunteer soldiers. The maneuver should be conducted by smaller but more talented and experienced units. Consequently, the military transformation aimed at building a more deadly maneuver force is smaller but more capable. This force structure is lighter on the ground, yet heavy in the air.[27]

Before closing this topic, one last point should be noted. As mentioned, units disperse, and the force-to-space ratio decreases at the operational level. The future will not see another Battle of Austerlitz. Even if mass armies are to be found, they could not be used efficiently due to space constraints. On the other hand, conventional battles are still pertinent. They now occur in narrow valleys or urban landscapes with huge opportunities for stealth. And like bees,[28] the modern system of fighters also concentrates its forces on these hot spots, with troop density increasing at the tactical level. Operational tempo is low, but intense tactical engagement and close-quarter battles are still the main characteristics of land battles. Contained offensives[29] can be said to be the new format of contemporary conventional battles.

Combined Arms: Coordinating the Fire and Maneuver Dimensions

Precision strike capability is necessary to secure a region and deny it to the adversary, but a modern system army is just as important as modern firepower. To reflect the full potential of combat power, these two dimensions should cooperate. In its simplest definition, this coordination process, called *combined arms*, is the practical reflection of combat power, and its critical component is the integration of firepower and maneuverability. Enhanced C4 systems and pivotal actors are needed in order to manage this process. These pivotal actors are the tactical air controllers (TACs).[30] The TACs' mission is to provide instantaneous firepower at any location, either independently or as a part of a maneuver unit. Thus, they act both as human sensors and command and control posts.[31]

The TACs' mission has become more critical against adversaries that apply the modern system. For example, targets may evade electronic sensors using exposure-reduction techniques such as natural and artificial terrain. In this scenario, one needs to get closer to the targets and coordinate firepower to hold or claim territory at the tactical level. These kinds of encounters were common in Afghanistan, Syria, and Iraq. As a result, several TACs were killed by enemy and friendly fire.

The Transformation of Türkiye's Warfare Capacity

How did the transformation of maneuver and firepower, as summarized above, take place in TAF? To answer this question, the change in TAF's organizational structure and doctrine are also examined alongside the defense industry projects and purchases carried out by Türkiye.[32] Chronologically, TAF's transformation in the post-1990 period took place in two phases. The catalyst for both phases has been the fight against PKK/KCK. In fact, this transformation impacted not only TAF but the whole Turkish security community as well. For this reason, this transformation includes the gendarmerie, police, and intelligence units that are actively involved in the fight against the PKK/KCK terrorist organization.

The First Stage

The Turkish Armed Forces underwent significant technical, tactical, and organizational changes during the first stage, which began in the early 1990s. The danger PKK/KCK posed acted as a catalyst for these reforms as TAF recognized the risk of losing territorial control in some regions of the country and responded.[33]

The Fire Dimension

One of the most important developments of this era was the transfer of NATO's excess military equipment in 1992 as a result of the Treaty on Conventional Forces in Europe. In this context, 1,050 M-60s and Leopard main battle tanks, 700 armored carriers, 40 F-4 fighters, and Cobra attack helicopters have been transferred to Türkiye.[34] In addition to these, much excess American armored equipment was also transferred to Türkiye after the Gulf War. Another important development was the joint production of F-16 fighter aircraft in Türkiye, and TAF has been using these fighters since 1987.

At this stage, TAF aimed to slow down the momentum of PKK/KCK's border campaign by creating a technological asymmetry. In this manner, border posts were first strengthened for defensive purposes. During this period, TAF's standard outposts were reinforced with mines and camouflaged positions and supported by firepower elements such as tanks, artilleries, and anti-aircraft guns, transforming them into fortified positions.[35] Second, several cross-border operations were carried out with an offensive spirit. These operations were carried out in the form of large-scale operations in which armored units, fighter jets, and tens of thousands of soldiers participated. Finally, TAF also started to own the night in this era. Since the early 1990s, the gradual acquisition of attack helicopters and tanks with thermal sight systems turned the night factor to TAF's advantage.

The Maneuver Dimension

The most significant change to TAF maneuver units came with the implementation of an area denial strategy. This strategy was built on the personnel-intensive, search-and-destroy tactics, which necessitated quantity over quality, resulting in the formation of many infantry units. This is consistent with the classical doctrine of counterinsurgency.[36] To raise more infantry units, the draft system was modified in 1994 by being extended from 15 to 18 months. In addition to this, lots of new gendarmerie units were formed from draft soldiers. As a result, gendarmerie units reached a force of over 200,000 and became an integral part of the area denial strategy. Furthermore, army and gendarmerie troops engaged in operations against the PKK/KCK started to receive intensive commando training.[37]

In the 1990s, the organizational structure and equipment of maneuver units changed, thanks to modernization projects. For example, the procurement of the UH-60 Black Hawk utility helicopters and Kobra armored vehicles provided protection and power projection capacity to the security forces.[38] In addition, night vision technology was also a force multiplier for the infantry units. By gaining control of the night, the resilience of military posts increased, and army units managed to claim 24 hours a day of area denial.[39]

During this time, the area denial accomplished by conscripted troops started to be supported by fully professional units. The Special Forces Command was formed as a joint and independent command. Additionally, special police operations units have been extensively utilized in the battle against the PKK/KCK to alleviate TAF's burden by participating in direct action operations in urban and rural areas.

The Second Stage

The second stage is based on improvements to the technological innovations of the 1990s and on adopting an all-volunteer force structure. Although societal factors and public pressure exist behind this decision, this chapter will only study the operational framework.

The Fire Dimension

Precision targeting technology has been the primary focus of the second phase. Türkiye has invested in its precision engagement capabilities since 2007, with precision guided munitions (PGMs) having been operational since 2010. Apart from the US-made Joint Direct Attack Missiles, Türkiye also uses locally manufactured PGM products from ASELSAN, ROKETSAN, and TÜBİTAK-SAGE.

The prominent platform in the development of precision strike capability has been uncrewed aerial vehicles (UAVs). Although TAF started its investments in UAV technology in 1994, this technology's contribution to organizational change did not become apparent until after 2010.[40] Türkiye has achieved a significant milestone in UAV technology since 2014, with a couple of models ready for mass production (i.e., the Aselsan ANKA and Bayraktar TB2). The transformation of TAF's area denial strategy and doctrine owes a lot to the TB2 UAVs. Thanks to their low cost and ability to be mass produced, the Bayraktar TB2 entered the inventory quickly and has been produced in vast numbers. As of February 2022, more than 100 UAVs (TB2-ANKA-Heron) were claimed to be in TAF's inventory.[41] Besides these combat-proven UAVs, advanced models such as the Akıncı and Aksungur have also been operationally ready as of late 2021.[42]

UAV technology has provided Türkiye with a valuable (ISR) capacity. The role of these platforms has dramatically increased TAF's ability to cover and control large areas like an electronic blanket. Accordingly, ANKA's and TB2's total flight times have increased more than 10-fold, from 36,894 hours in 2017 to upwards of 400,000 as of November 2021.[43] The number of UAV bases in Türkiye has also steadily increased. These bases not only cover eastern Türkiye but also northern Iraq and northern Syria.

Türkiye's UAV capability is not confined to the new area denial strategy but has also contributed significantly to targeted operations and close air support missions conducted abroad. They have been used in the Euphrates Shield, Olive Branch, Peace Spring, and Claw operations. They are also actively used by expeditionary forces in Libya. Türkiye has used its UAV capacity in innovative ways as well. The air interdiction campaign against Syrian ground forces in Operation Spring Shield under heavy anti-access/area denial (A2AD) threat can be listed as an innovative approach, as well as the idea of designing the TCG Anadolu as a drone carrier ship. UAV footage has also been used as an information warfare tool against PKK/KCK claims of collateral damage.[44]

In addition to UAVs, TAF's attack helicopter capacity has also multiplied since 2014. Infantry units had to rely on a limited number of these helicopters during the 1990s, but Türkiye now possesses the robust firepower support of attack helicopters developed under the T-129 ATAK program. As of August 2021, TAF has 56 T-129 attack helicopters in its inventory.[45]

These firepower elements are not limited to the armed forces. Law enforcement and intelligence agencies also have massive fire support in their organic structure. As of September 2021, gendarmerie forces have nine and police forces another three T-129 attack helicopters.[46] These numbers will reach 24 and nine, respectively. Their UAV capacity has also been enhanced. As of November 2020, gendarmerie forces possess 45 UAVs (39 TB2s and six Anka-Ss), and police forces have 19.[47] Most of these UAVs have PGM capacity. Another critical point involves the intelligence agency (*Milli İstihbarat Teşkilatı* [MİT]) which owns 14 UAVs (Both TB2s and Ankas). With its modern firepower capacity, MİT has become an integral actor of targeted operations.[48]

Last but not least, the Network Centric Warfare (NCW) model has also become a component of TAF's transformation process.[49] Airborne warning and control systems (AWACS) and other crewed and uncrewed C4ISR platforms, Link16 infrastructure, and electronic warfare capabilities are valuable force multipliers in NCW. This NCW model proved its effectiveness in Operation Peace Spring against Syrian armed forces in a contested A2AD environment.

The Maneuver Dimension

Modern firepower technology has also triggered transformation in TAF's organizational structure and changed the area denial tactics of the 1990s. As stated in the first stage, TAF had tried to establish area denial through mass troops in the 1990s, using the manpower-intensive find-fix-fight doctrine. On the contrary, TAF once enabled with precision targeting technology has progressively passed to an electronic area denial doctrine and terminated the mass conscripted force-based counterterrorism strategy. Consequently, the transformed doctrine of area denial allowed TAF to introduce a professional modern army system with sufficient training, tactical knowledge, and experience.

Although the offspring of professionalization has matured in this second decade of the 21st century, the starting point of the transformation efforts can be dated back to the mid-1980s. In this manner, the initial point of professionalization was the recruitment of specialist sergeants in 1986. The aim was to assign expert cadres to positions such as tank driver, radio operator, and radar operator. The second stage of the transformation began in 2001 with the professionalization of infantry commando brigades. TAF planned to fulfill its low-level command cadres by recruiting contracted officers and petty officers. The third phase was projected to professionalize all the commando units of the army. By 2010 the project was accomplished, with five infantry and one gendarmerie commando brigade having been transformed into all-volunteer units. The last stage of the professionalization of the maneuver units was the recruitment of contracted privates in 2011.[50]

These projects accelerated TAF's transformation, and the number of commando brigades dramatically increased during the second decade of the 21st century. Infantry brigades in Denizli, Sarıkamış, Tatvan, and Bingöl were transformed into commando brigades in 2016.[51] Moreover, new commando brigades formed in Sakarya, Trabzon, and Kırklareli in 2018. Commando battalions have additionally been formed within the organic structure of mechanized infantry brigades. The amphibious brigade in İzmir has also been actively used in combat zones in Iraq and Syria. By 2021, TAF had 17 all-volunteer commando brigades.[52] Compared to the force structure of the 1990s, TAF has made a considerable capacity increase in the context of *modern system force employment*. Based on these changes, one might conclude that, in contrast to the 1990s, commando missions are no longer a special duty for TAF units. Commando brigades are the infantry of the 21st century's modern warfare.

Ground units' maneuvering equipment has also been continuously modernized. The mine-resistant ambush protected (MRAP) variants of the Otokar Kobra, Nurol Ejder, and BMC Kirpi are now standard vehicles. The aviation capacity of security forces is also constantly improving. The number of total modern utility helicopters that are used by TAF, police, and gendarmerie forces is expected to reach up to 400 in this decade.[53]

Professional units of the gendarmerie and police forces have also been an integral part of the transformation. As of 2016, the number of gendarmerie[54] and police[55] special operation units were 12,500 and 7,800, respectively. These numbers are planned to reach around 20,000 in both services.[56] Gendarmerie and police special operation units have nearly the same capacity as infantry commando units. They train, equip, and fight as a modern army system. They are used not only in targeted direct-action missions but also as maneuver elements in operations both in Türkiye and abroad. For example, these units were extensively used against fortified PKK/KCK positions in Operation Olive Branch.

The all-volunteer model has also enhanced the combined arms of fire and maneuver. One development in this manner has been the frequent employment of TACs. Although TAF used airmen as TACs against PKK/KCK targets in the first stage of its transformation, targets were fixed in traditional ways due to the lack of any laser designators. However, Aselsan's production of laser designators since 2009 has been effectively used by TACs.[57] Currently, these devices are frequently used not only by special forces but also by conventional commando units and forward bases.

Analysis of Türkiye's Warfare Capability

The critical turning point for the TAF transformation process was 2012, when the outcomes of the transformation became visible during comprehensive military operations against PKK/KCK, ISIS, and the Syrian Army.[58] Naturally, the technological edge, professional units, modern system maneuvers, and joint operations all played important roles in these operations. However, this transformation also created a response, and the target actors have modified their strategies and tactics. This action-reaction relationship can be traced by examining PKK/KCK's transformation process. Thus, 2012 was ironically also an important year for PKK/KCK's transformation.

Action: A New Doctrine for an Old Problem

It can be said that TAF has successfully integrated its new firepower and maneuver capacity into its doctrine. Accordingly, for the first time in history, TAF conducted a conventional operation in 2012 using only an all-volunteer professional force structure. Since then, professional units have conducted all these comprehensive operations. Another significant result of the new doctrine is the changing character of the area denial strategy. TAF changed its infantry-based doctrine into a technology-based one. In the absence of hundreds of thousands of draft soldiers, intelligence has become the main weapon of the area denial strategy. Operations are no longer planned according to troop density but instead according to technological edge. In major operations, PGMs and TACs are heavily used in coordination with maneuver units. All-volunteer professional units allow operations to be conducted with fewer and fewer troops (see Figure 7.1). For example, the mean number of troops used in five comprehensive operations against PKK/KCK bases in Iraq between 1992 and 2008 was 36,000. Since 2012, this number is now less than 7,000.[59] This is one of the concrete results of TAF's transformation. Personnel has finally been replaced by technology, and the modern system maneuver has become the standard.[60]

In this context, Operation Dawn in 2012 is an essential reference point for TAF's new approach. The operation was planned against a crowded PKK/KCK group in Şemdinli that had been trying to hold a specific ground in fortified positions; the operation lasted for 11 days between September 8 and 19, 2012. PKK/KCK attempted to exploit the terrain to conceal and camouflage their defensive positions. Operation Dawn started with heavy close air support to detect and break down the defensive positions. In this phase, approximately ten TACs were actively used to locate the concealed positions. In the second phase of the operation, the maneuver units engaged the dispersed groups of PKK/KCK. The number of these maneuver forces was approximately 2,000, and they were composed of infantry commandoes, special forces, gendarmerie, police special forces, and local village guards. During the maneuver phase of the operation, PKK/KCK persistently tried to hold

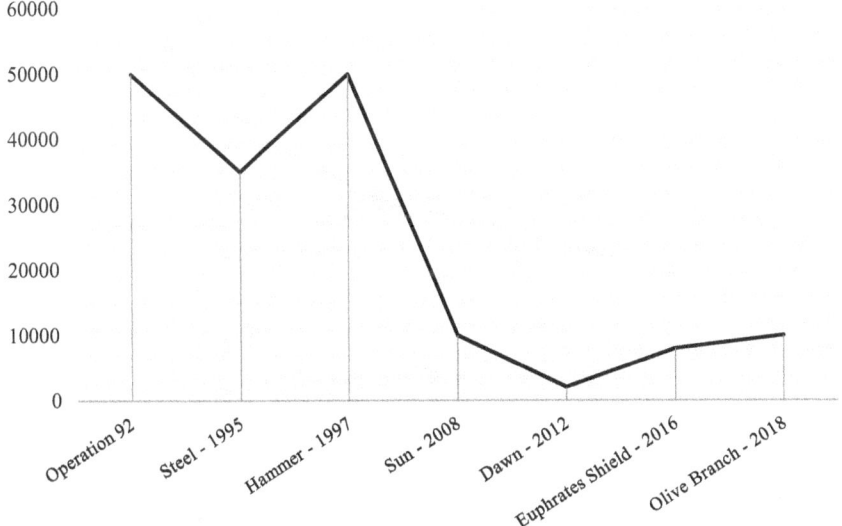

Figure 7.1 Number of Security Forces Used.

Sources: *Created by the author using open-source records.*

its positions, deployed reserve units, and made counterattacks. At the end of the operation, 137 PKK/KCK members had been neutralized, with one captured alive. Security force casualties numbered eight killed in action and nine slightly injured during the maneuver phase of the operation.[61] This ratio of 15/1 can be seen as a reasonable casualty ratio in this type of conventional engagement. Moreover, the number of wounded personnel was remarkably low. For comparison, in Operation Anaconda, conducted in 2002 in Afghanistan, US casualties totaled eight killed in action and over 50 wounded, while the casualties from the Afghan proxies were much higher.[62]

As of 2022, TAF continues its cross-border operations in Iraq. The goal of the Claw Operations is to take physical control of the mountain ranges that PKK/KCK groups have heavily fortified. As the territory gives many opportunities for concealment and camouflage, the technological area-denial doctrine faces a challenge, and conventional maneuvers become the only way of holding an area. PKK/KCK knows this and tries to make this harder by practicing intense A2AD measures such as tunnels and improvised explosive devices (IEDs). The Claw Operations are conducted gradually due to the rugged terrain, harsh weather conditions, logistical constraints, and heavy A2AD threat. A new range of hills has been captured and fortified by maneuver units at each stage. Although ISR is heavily used, the main actor is the infantry, both offensively and defensively. As a result, PKK/KCK is gradually losing ground to TAF and getting pushed back to the south.[63] Accordingly, the geographical center of gravity of the clashes has transited from Türkiye to northern Iraq (see Figure 7.2).

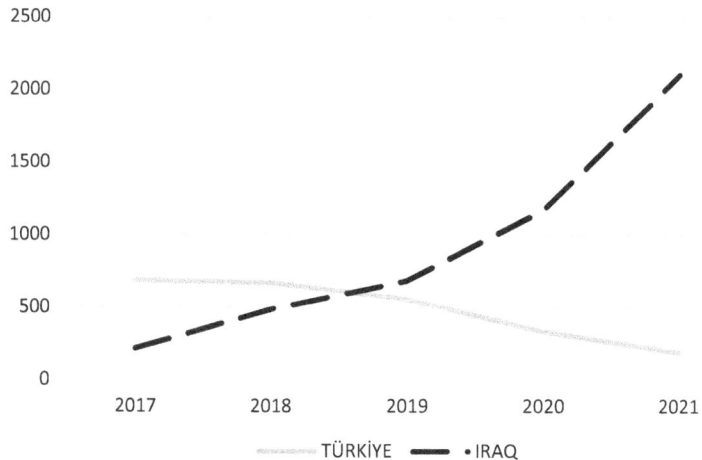

Figure 7.2 Number of Clashes Between Security Organizations and PKK/KCK.

Sources: *Created by the author. Source: Armed Conflict Location & Event Data Project (ACLED) Codebook.*

The author used the data from the Armed Conflict Location & Event Data Project (ACLED) Codebook to create the figure. "Armed clashes", "air/drone strikes" and "shelling/artillery/missile attacks" are included in the statistics. For similar statistical results, see also Adam Miller, Turkey-PKK Conflict: Rising Violence in Northern Iraq, https://acleddata.com/2022/02/03/turkey-pkk-conflict-rising-violence-in-northern-iraq/, accessed on 10 June 2022.

Response: PKK/KCK's Adaptation

The evidence of TAF's transformation can also be traced by analyzing the transformation of the narrative and modus operandi of PKK/KCK. A chronological overlap exists between the two transformation processes. The driving force behind PKK/KCK's transformation has been to offset the technological asymmetry.

First of all, PKK/KCK responded to the time dimension by ceasing its night attacks. Because TAF owns the night, PKK/KCK tried to exploit daylight opportunities. This response can be traced statistically as well. Between August 2012 and October 2018, 175 of 182 attacks (96% of all attacks) were conducted in daylight.[64] PKK/KCK's second response was in its organizational structure. They are seen to have been attempting to use modern system force employment tactics since at least 2012. In this context, their regiment and battalion formations have been terminated, with team and small team formations being formed to conduct *small-unit maneuvers*. Moreover, *concealment* and *camouflage* have become the highest priorities. In PKK/KCK's narrative, this is called the *modern professional guerilla of the 21st century*.

Ironically, these modern system tactics have resulted in some complications at the operational level. In 2012, PKK/KCK lost its full-scale conventional attack capability[65] to TAF military outposts.[66] After 30 years of a heavy border campaign, PKK/KCK suddenly ceased its attacks and turned toward a defensive posture. Since then, the modus operandi has been sabotage, harassment, small unit seizure at border posts, and, most importantly, forming conventional defenses.

This change can be seen by analyzing the frequency of border attacks in the Hakkari province. During the 1990s, PKK/KCK's primary strategy was to gain territory through conventional attacks. In 1991 and 1992 in particular, they tried to drive TAF from the Iraqi border after a series of attacks. The strategy was annihilation. Through three attacks in just one month between August 30 and September 29, 1992, TAF suffered 73 casualties. These attacks were conducted by large maneuvering groups (some of them up to 500 strong) under firepower protection (e.g., RPGs, anti-aircraft guns, and mortars). Due to the heavy PKK/KCK losses and TAF's cross-border operations, PKK/KCK's border campaign strategy inevitably turned from annihilation to attrition between 1993 and 2012. However, the total number of casualties was still huge. TAF suffered 262 losses from 26 conventional attacks on the Hakkari border during this period, which constitutes nearly 20% of all casualties.[67]

As was mentioned, PKK/KCK had managed to protect its conventional attack capacity until 2012. On the other hand, no conventional attacks occurred between 2013 and 2022. This situation chronologically matches TAF's transformation process. Technological asymmetry had reduced PKK/KCK's conventional attack capacity and will. This inevitably caused them to alter their strategy into a defensive one. PKK/KCK wanted to confront TAF in fortified defensive positions in the urban and mountainous landscape to offset the technological gap with its new defensive strategy. Searching for maneuvering targets was the main reason for the change in PKK/KCK's strategy. The leadership of the organization frequently stressed that, because of the technological gap, they should search for attacking (maneuvering) targets, not defending ones. This was the driving force behind the 2012 Şemdinli plan, the 2015–2016 urban conflicts in Sur, Nusaybin, and Cizre, and the 2018 Afrin conflict.[68] The aim was to utilize the natural and artificial terrain for stealth and to concentrate their forces at decisive points. Although these plans did not work and PKK/KCK suffered huge casualties, they did manage to compel security forces to maneuver in rugged landscapes.[69]

Conclusion

This chapter has explained Türkiye's military transformation under the theoretical framework of firepower (precision engagement) and maneuverability (modern system of force employment). The main aim of this transformation was to form a force structure that is light on land and heavy in the air. With its modern firepower capacity and professional maneuver units, TAF has transformed into a modern system force structure.

TAF has been investing in these capabilities since the beginning of the 21st century. In this manner, it has acquired various kinds of PGMs and UAVs. Moreover, it has transformed its maneuvering units into a volunteer structure. However, this transformation has not just been limited to TAF. Other Turkish security agencies, gendarmerie, police forces, and intelligence agencies have also transformed themselves.

The footsteps of this transformation process can be followed through the transformation doctrines. TAF changed its doctrine from personnel-intense, search-and-destroy campaigns into an intelligence-driven, find-fix-fight concept. In this new doctrine, area denial is achieved not by quantity but through technology, and conventional engagements are conducted by small, agile volunteer forces supported by precision munitions.

The transformation has also forced PKK/KCK to change their concept. Their main motivation was to offset the technological asymmetry. As a result, PKK/KCK deliberately started to implement *small unit independent maneuver* and *exposure-reduction tactics* such as *concealment, cover*, and *camouflage*. On the other hand, this caused PKK/KCK to lose its ability to organize conventional attacks and inevitably forced it to accept a defensive posture. In conclusion, to say that TAF's transformation in the post-Cold War era has accomplished its objectives, changed the character of conflict, and caused disastrous results for PKK/KCK would be a fair assessment.

Notes

1 The armed wing of PKK/KCK is HPG (Formerly HRK and ARGK). To simplify, only the abbreviation PKK/KCK is used.
2 See Geoffrey Parker, "The Military Revolution 1560–1660 – A Myth?" *Journal of Modern History*, 48, 1976, pp. 195–214.

3 See Nikolai Vasilyevich Ogarkov, "The Defense of Socialism: Experience of History and the Present Day", *Red Star*, May 9, 1984 cited in Barry Watts, *The Evolution of Precision Strike*, Center for Strategic and Budgetary Assessments, Washington, 2013, p. 6.

4 See Andrew Krepinevich and Barry Watts, *The Last Warrior*, Basic Books, New York, 2015.

5 One of the main reasons for this terminological change was the big expectations the term "revolution" had created. That is why they wanted to use a more modest term like "transformation". On the other hand, "transformation" is broader than "reform" or "modernization" and symbolizes much deeper changes. See Elinor Sloan, *Military Transformation and Modern Warfare A Reference Handbook*, Westport: Praeger, 2008, p. 7–8; Barış Ateş, "Military Change: An Analysis on the Post-Cold War Era", *Çankırı Karatekin University Journal of the Faculty of Economics and Administrative Sciences*, 10:1, p. 18.

6 Donald H. Rumsfeld, "Transforming the Military", *Foreign Affairs*, 2002, 81:3, p. 29.

7 In the Joint Vision 2020 document, the combat power is studied under the title of *Full Spectrum Dominance*. *Full Spectrum Dominance* is composed of *dominant maneuver, precision engagement, focused logistics,* and *full-dimensional protection*. These four dimensions form the combat power of a military entity. See Joint Vision 2020. *America's Military – Preparing for Tomorrow*, Institute for National Strategic Studies, Washington, 2000.

8 Keir A. Lieber, "Grasping the Technological Peace the Offense-Defense Balance and International Security", *International Security*, 25:2, 2000, p. 80.

9 Although naval-based firepower has been used for naval warfare, coercive bombardment of coastal cities, and supporting amphibious maneuvers, these kinds of concepts are not involved in this chapter.

10 In JSOC terminology this sequence is called F3EA (Find-Fix-Finish-Exploit-Analyze). In this concept, finishing is a kill/capture mission and it can be done by special forces. After finishing a target, the exploited and analyzed pieces of evidence start another finding mission. That is why this is a cycle rather than a sequence. See Stanley McChrystal, "It Takes a Network: The New Frontline of Modern Warfare", *Foreign Policy*, 2011. https://foreignpolicy.com/2011/02/21/it-takes-a-network/ accessed on 10 June 2020.

11 Karl P. Mueller, "Examining the Air Campaign in Libya", in (eds.) Karl P. Mueller, *Precision and Purpose: Airpower in the Libyan Civil War*, RAND, Santa Monica, 2015, p. 4.

12 The kill chains were three days during the Gulf War. Today it is calculated by minutes due to enhanced C4ISR technology. See Max Boot, "The New American Way of War", *Foreign Affairs*, 82:4, 2003, p. 52.

13 Michael Warner, "Reflections on Technology and Intelligence Systems", *Intelligence and National Security*, 27:1, 2012, p. 151.

14 Joint Publication 3-0 Joint Operations, *Joint Chiefs of Staff*, Washington, 2017, p. xiv.

15 Biddle classifies this aim as Napoleonic. See Stephen Biddle, *Nonstate Warfare: The Military Methods of Guerillas, Warlords, and Militias*, Princeton University Press, New Jersey, 2021.

16 Max, Boot, *Invisible Armies: An Epic History of Guerrilla Warfare from Ancient Times to the Present*, Liveright, 2013, p. 56.

17 Williamson Murray, "Thinking About Revolutions in Military Affairs", *Joint Force Quarterly*, 16, 1997, p. 72.

18 For Germany's WW1 infantry tactics see Anthony King, *The Combat Soldier Infantry Tactics and Cohesion in the Twentieth and Twenty-First Centuries*, Oxford University Press, 2013, p. 133; Jonathan M. House, *Toward Combined Arms*

Warfare: A Survey of 20th Century Tactics, Doctrine, and Organization, Combat Studies Institute, Kansas, 1984, p. 34–35.

19 House, *Toward Combined Arms Warfare*, p. 36.
20 See Biddle, *Nonstate Warfare*, chapter 3.
21 Gordon R. Sullivan and James M. Dubik, *Land Warfare in the 21st Century*, US Army War College Strategic Studies Institute, Washington, 1993, p. 16 and Trevor Dupuy, *The Evolution of Weapons and Warfare*, The Bobbs-Merrill Company, 1980, p. 312.
22 Michael G. Vickers and Robert C. Martinage, *The Revolution in War*, Center for Strategic and Budgetary Assessments, Washington, 2004, p. 169.
23 Biddle, *Nonstate Warfare*, Chapters 3 and 4.
24 In the 2006 Lebanon War and in 2015–2016 urban clashes between Turkish security forces and PKK/KCK, both Israeli and Turkish reserve units had casualties due to the heavy sniper fire.
25 Stephen Biddle, *Military Power Explaining Victory and Defeat in Modern Battle*, Princeton University Press, New Jersey, 2004, p. 28, 44.
26 Stephen Biddle, "Rebuilding the Foundations of Offense-Defense Theory", *The Journal of Politics*, 63:3, 2001, p. 755.
27 Kenneth H. Watman and Daniel P. Raymer, *Airpower in US Light Combat Operations*, RAND, Santa Monica, 1994, p. 14.
28 The "Bee metaphor" was also used by Arquilla and Ronfeldt in a slightly different context to describe the battle swarm model. For more information see John Arquilla and David Ronfeldt, *Swarming and the Future of Conflict*, RAND, Santa Monica, 2000, p. 25.
29 Biddle uses *contained offensive* to identify modern to modern system confrontations. See *Military Power*, p. 74.
30 In manuals, there are different classifications and terminologies for air controller missions. See Joint Publication 3-09.3 Close Air Support, *Joint Chiefs of Staff*, Washington, 2014. For detailed information on air controller missions see also Bruce R. Pirnie et al., *Beyond Close Air Support Forging a New Air-Ground Partnership*, RAND, Santa Monica, 2005. For details of laser designators see also JP 3-09.3.
31 For more information see United States Special Operations Command Fact Book, USSOCOM Office of Communication, 2017.
32 The technical details of defense contracts are not mentioned in this chapter.
33 Kasapoğlu calls this transformation Strategy of 1993. See Can Kasapoğlu, Düşük Yoğunluklu Çatışmalarda Konvansiyonel Güçlerin Kullanılması: 1991–1999 Türk Silahlı Kuvvetleri Örneği, unpublished doctorate thesis, Harp Akademileri Komutanlığı Stratejik Araştırmalar Enstitüsü, İstanbul, 2011.
34 Cenk Özgen, "An Example of Conversion in the Security Field from 1980s to 2000s: Treatment of the Legal Arrangements Regarding Professional Military Service in Mainstream Journals", *Uludağ University, Social Sciences Journal*, 14:24, 2013, p. 102.
35 See Osman Pamukoğlu, *Unutulanlar Dışında Yeni Bir Şey Yok*, İnkılap Kitapevi, İstanbul, 2004, p. 28.
36 Troop density in counterinsurgency campaigns is an important debate. There are some mathematical formulas too. For example, according to FM 3–24, the preponderance of counterinsurgency units should be 10–15 to 1 against the armed insurgents and there should be 20–25 soldiers per 1000 civilians. See Field Manual 3-24 / Marine Corps Warfighting Publication No. 3–33.5, University of Chicago Press, Chicago, 2007, pp. 22–23. On the other hand, Friedman claims that according to his research on 171 counterinsurgency campaigns there is not a valid rule of thumb like 20–25/1000 troop density, and the successful campaigns generally have a ratio of 80/1000. See Jeffrey A. Friedman, "Manpower and Counterinsurgency: Empirical Foundations for Theory and Doctrine", *Security Studies*. 20:4, 2011, p. 581–583.

37 In 1993 in Hakkari Mountain Commando Brigade only 11 of the 250 officers had completed the commando course, Pamukoğlu, *Unutulanlar Dışında*, p. 106.

38 The first of these Black Hawks was assigned to the gendarmerie in 1989, "Türk Silahlı Kuvvetleri ve Genel Maksat Helikopterleri (2)", https://www.savunmasanayist.com/turk-silahli-kuvvetleri-ve-genel-maksat-helikopterleri-2/ accessed on 30 November 2021.

39 Interview with Doğan Güreş, Fikret Bila, *Komutanlar Cephesi*, Detay Yayıncılık, İstanbul, 2007, p. 48.

40 For a chronology of Türkiye's UAV journey, see İbrahim Sünnetçi, "İHA'lar ve Türkiye'nin İnsansız Havadan İstihbarat Çalışmaları", *Savunma ve Havacılık Dergisi*, 132, 2009, pp. 75–80.

41 "Türkiye'nin SİHA gücü ne kadar? Hangi modellerde kaç tane var? Çatışmalarda hangileri tercih ediliyor? İşte cevapları" https://www.indyturk.com/node/342926/haber/t%C3%BCrkiyenin-si%CC%87ha-%C3%BCc%C3%BC-ne-kadar-hangi-modellerde-ka%C3%A7-tane-var-%C3%A7at%C4%B1%C5%9Fmalarda accessed on 05 April 2021.

42 Three Akıncı and one Aksungur UAV were delivered to TAF in 2021.

43 "Türkiye'nin milli silahlı hava aracı Bayraktar TB2'ler rekor kırdı", https://www.trthaber.com/foto-galeri/turkiyenin-milli-silahli-hava-araci-bayraktar-tb2ler-rekor-kirdi/40860/sayfa-5.html accessed on 26 November 2021; ANKA Multirole UAV System https://www.tusas.com/content/files/uploads/2232/TUSAS_2020_Genel_Flyer_Anka_EN.pdf accessed on 26 November 2021; 100000 hours for gendarmerie Twitter account of Gendarmerie General Command accessed on 26 November 2021.

44 For Information Warfare aspect see Dağhan Yet's chapter in this book.

45 "Kara Kuvvetleri'nin ATAK helikopter sayısı 56'ya ulaştı", https://haber.aero/aero-gundem/t129-atak-helikopter-orduya-teslim-edildi/ accessed on 01 April 2022; in the ATAK Programme, a total of 91 (59 agreed 32 optioned) T-129 is planned to be produced for TAF. See Turkish Aerospace official web page, https://www.tusas.com/en/products/helicopter/co-development-production/t129-atak accessed on 01 Nov 2021.

46 Jandarma'ya yeni T129 ATAK FAZ-2 taarruz helikopteri teslimatı https://www.defenceturk.net/jandarmaya-yeni-t-129-atak-faz-2-taarruz-helikopteri-teslimati, accessed on 25 November 2021. In the ATAK Program, a total of 27 (23 agreed 3 optioned) T-129s are planned to be produced for gendarmerie and police forces. See Turkish Aerospace official web page.

47 General Directorate of Security 2020 Activity Report, https://www.egm.gov.tr/kurumlar/egm.gov.tr/IcSite/strateji/Planlama/2020_IDARE_FAALIYET_RAPORU.pdfp.8, accessed on 22 November 2021.

48 Traditionally intelligence agencies are expected to find the targets and pass the intelligence to the operational units. On the other hand, MİT transformed into an operational actor with its fighting capacity, too.

49 See Arthur K. Cebrowski, and John H. Garska, "Network-Centric Warfare: Its Origin and Future", *US Naval Institute Proceedings Magazine*, 1998, http://www.iwar.org.uk/rma/resources/ncw/network-centric-warfare.ht accessed 07 June 2018.

50 For more info see Cenk Özgen, *"Professionalization Tryouts in Turkish Armed Forces"*, *Trakya University Social Science Journal*, 13:1, 2011, p. 200; Özgen, 2013, p. 102.

51 "PKK'nın değişen taktiklerine karşı yeni strateji", http://www.hurriyet.com.tr/gundem/pkknin-degisen-taktiklerine-karsi-yeni-strateji-40094019, accessed on 02 October 2021.

52 "Hulusi Akar TSK'nın başarı bilançosunu anlattı: 2 komando tugayı vardı, şimdi 17", https://www.hurriyet.com.tr/gundem/hulusi-akar-tsknin-basari-bilancosunu-anlatti-2-komando-tugayi-vardi-simdi-17-41835653, accessed on 27 June 2021.

53 Currently, there are 106 UH-60 Black Hawks, 48 AS532 Cougars, 17 Mi-17s, and 11 CH-47 Chinooks in the inventory. The retro Bell UH-1 Iroquois are planned to be called out of service in the forthcoming years. In the 2020s, 219 (121 agreed 98 optioned) more UH-60 Black Hawks (six of them for the navy) will be added to the inventory. In addition to them, the project of T-625 Gökbey is also on its way.

54 As of 2020, in the gendarmerie forces, there are 155.091 professional cadres. Gendarmerie General Command 2020 Activity Report, p. 11, https://www.jandarma.gov.tr/jandarma-genel-komutanligi-2020-yili-faaliyet-raporu, accessed on 27 December 2021.

55 As of 2020 in the police forces, there are 328,963 professional cadres. General Directorate of Security 2020 Activity Report, p. 14.

56 "Özel Harekatçı sayısı 40 bine çıkıyor", http://www.hurriyet.com.tr/gundem/ozel-harekatci-sayisi-40-bine-cikiyor-40071325, accessed on 14 June 2020.

57 Aselsan-ENGEREK Lazer Mesafe Ölçme Cihazı- Laser Target Designator/Locator, https://www.youtube.com/watch?v=ScorhAk1XDM, accessed on 16 March 2020.

58 According to Minister of Defense Hulusi Akar, TAF conducted 13 comprehensive military operations between 2016 and 2021 "Hulusi Akar TSK'nın başarı bilançosunu anlattı".

59 Operation Euphrates Shied was conducted against ISIL.

60 Operation Dawn in many ways resembles Operation Anaconda, which was carried out by US forces in Afghanistan. On the other hand, it can be seen that Operation Dawn was much more carefully planned and successfully executed.

61 "Şemdinli Şafak Operasyonu", https://www.hurriyet.com.tr/gundem/semdinli-safak-operasyonu-21563693, accessed on 15 March 2020.

62 Richard L. Kugler, *Operation Anaconda in Afghanistan: A Case Study of Adaptation in Battle*, National Defense University, Washington, 2007, p. 1.

63 This is a slow-moving gradual process which Biddle called *contained offensive*. See Biddle, *Military Power*, p. 74.

64 Numbers are derived from open sources and only the recorded attacks are counted.

65 Biddle and Friedman's scales are used in the conceptualization of a conventional engagement. These scales are briefly: *the duration of firefights, the proximity of attackers to defenders*, and *the incidence of counterattacks*. An increase in these variables reflects the intention of *holding ground, large-scale maneuver with combined arms*, and *decisive engagement*. In the conventional approach, as opposed to unconventional warfare, force density increases at decisive points, and more detailed planning and coordination are required at the operational level in the means of firepower, maneuver, logistics, and communication. See Stephen Biddle and Jeffrey A. Friedman, *The 2006 Lebanon Campaign and the Future of Warfare: Implications for Army and Defense Policy*, Strategic Studies Institute, Pennsylvania, 2008; Biddle also calls this the Fabian-Napoleonic continuum of warfare. See Biddle, *Nonstate Warfare*; Mearsheimer classifies conventional engagements as the blitzkrieg, attrition and limited aims strategy. See John J. Mearsheimer, *Conventional Deterrence*, Cornell Uni. Press, Ithaca, 1983, pp. 23–66.

66 PKK/KCK's last conventional attack was toward the Geçimli outpost in the Hakkari region of Türkiye. A group of approximately 50 PKK/KCK members conducted this attack and eight soldiers were killed in action. "Geçimli geçilmedi: 8 şehit, 14 PKK'lı öldürüldü", https://www.hurriyet.com.tr/gundem/gecimli-gecilmedi-8-sehit-14-pkk-li-olduruldu-21158385,accessed on 17 June 2021.

67 Numbers are derived from open sources.

68 This change is consistent with the asymmetrical conflicts of the 21st century. Due to technological asymmetry, the weaker side of the conflicts tries to establish

conventional defense positions in both urban and rural decisive points. For example, in the 2006 Lebanon War, Hezbollah successfully lured Israel into maneuver warfare and overcame Israeli Defense Forces in subsequent tactical confrontations. Biddle and Friedman, *The 2006 Lebanon Campaign*, p. 77.
69 Security forces lost 71 personnel in Sur and 52 in Afrin operation. For example, during Operation Olive Branch on March 1st of 2018, a PKK/KCK group counterattacked and outflanked a maneuvering gendarmerie special forces unit by using preconstructed tunnels. Due to heavy fog and proximity of the engagement, close air support could not have been used effectively. The engagement lasted for 16 hours, and the gendarmerie unit had nine casualties. "Afrin Kahramanları Anlatıyor belgeseli 2. Bölüm", https://www.youtube.com/watch?v=0-oT5tnYIk4, accessed on 24 June 2021; This method resembles the "limited aims strategy" of Mearsheimer's classification. In this method, after initially seizing on a weak point by using the surprise factor, the aggressor seeks a battle of attrition in fortified defensive positions. For more information, see Mearsheimer, *Conventional Deterrence*, p. 53.

References

"Afrin Kahramanları Anlatıyor belgeseli 2. Bölüm", https://www.youtube.com/watch?v=0-oT5tnYIk4, accessed on 24 June 2021.

ANKA Multirole UAV System, https://www.tusas.com/content/files/uploads/2232/TUSAS_2020_Genel_Flyer_Anka_EN.pdf accessed on 26 November 2021.

Arquilla, John and Ronfeldt, David, *Swarming and the Future of Conflict*, RAND, Santa Monica, 2000.

"Aselsan | ENGEREK Lazer Mesafe Ölçme Cihazı - Laser Target Designator / Locator", https://www.youtube.com/watch?v=ScorhAk1XDM accessed on 16 March 2020.

"ATAK Programme", https://www.tusas.com/en/products/helicopter/co-development-production/t129-atak, accessed on 01 November 2021.

Ateş, Barış, "Military Change: An Analysis on the Post-Cold War Era", *Çankırı Karatekin University Journal of the Faculty of Economics and Administrative Sciences*, 10:1, pp. 15–42.

Biddle, Stephen, "Rebuilding the Foundations of Offense-Defense Theory", *The Journal of Politics*, 63:3, 2001, pp. 741–774.

Biddle, Stephen, *Military Power Explaining Victory and Defeat in Modern Battle*, Princeton University Press, New Jersey, 2004.

Biddle, Stephen, *Nonstate Warfare the Military Methods of Guerillas, Warlords, and Militias*, Princeton University Press, New Jersey, 2021.

Biddle, Stephen and Friedman, Jeffrey A., *The 2006 Lebanon Campaign and the Future of Warfare: Implications for Army and Defense Policy*, Strategic Studies Institute, Pennsylvania, 2008.

Bila, Fikret, *Komutanlar Cephesi*, Detay Yayıncılık, İstanbul, 2007.

Boot, Max, "The New American Way of War", *Foreign Affairs*, 82:4, 2003, pp. 41–58.

Boot, Max, *Invisible Armies: An Epic History of Guerrilla Warfare from Ancient Times to the Present*, Liveright, New York, 2013.

Cebrowski, Arthur K. and Garska, John H., "Network-Centric Warfare: Its Origin and Future", *US Naval Institute Proceedings Magazine*, 1998, http://www.iwar.org.uk/rma/resources/ncw/network-centric-warfare.ht accessed 07 June 2018.

Dupuy, Trevor, *The Evolution of Weapons and Warfare*, The Bobbs-Merrill Company, Indianapolis, 1980.

Field Manual 3-24 / Marine Corps Warfighting Publication No. 3-33.5, University of Chicago Press, Chicago, 2007.

Friedman, Jeffrey A., "Manpower and Counterinsurgency: Empirical Foundations for Theory and Doctrine", *Security Studies*. 20:4, 2011, pp. 556–591.

"Geçimli geçilmedi: 8 şehit, 14 PKK'lı öldürüldü", https://www.hurriyet.com.tr/ gundem/gecimli-gecilmedi-8-sehit-14-pkk-li-olduruldu-21158385 accessed on 17 June 2021.

Gendarmerie General Command 2020 Activity Report, p. 11, https://www.jandarma. gov.tr/jandarma-genel-komutanligi-2020-yili-faaliyet-raporu accessed on 27 December 2021.

General Directorate of Security 2020 Activity Report, https://www.egm.gov.tr/ kurumlar/egm.gov.tr/IcSite/strateji/Planlama/2020_IDARE_FAALIYET_ RAPORU.pdfp.8 accessed on 22 November 2021.

House, Jonathan M., *Toward Combined Arms Warfare: A Survey of 20th Century Tactics, Doctrine, and Organization*, Combat Studies Institute, Kansas, 1984.

"Hulusi Akar TSK'nın başarı bilançosunu anlattı: 2 komando tugayı vardı, şimdi 17", https://www.hurriyet.com.tr/gundem/hulusi-akar-tsknin-basari-bilancosunu-anlatti-2-komando-tugayi-vardi-simdi-17-41835653 accessed on 27 June 2021.

"Jandarma'ya yeni T129 ATAK FAZ-2 taarruz helikopteri teslimatı", https://www. defenceturk.net/jandarmaya-yeni-t-129-atak-faz-2-taarruz-helikopteri-teslimati accessed on 25 November 2021.

Joint Publication 3-0 Joint Operations, *Joint Chiefs of Staff*, Washington, 2017.

Joint Publication 3-09.3 Close Air Support, *Joint Chiefs of Staff*, Washington, 2014.

Joint Vision 2020. *America's Military - Preparing for Tomorrow*, Institute for National Strategic Studies, Washington, 2000.

"Kara Kuvvetleri'nin ATAK helikopter sayısı 56'ya ulaştı", https://haber.aero/ aero-gundem/t129-atak-helikopter-orduya-teslim-edildi/ accessed on 01 April 2022;

Kasapoğlu, Can, Düşük Yoğunluklu Çatışmalarda Konvansiyonel Güçlerin Kullanılması: 1991–1999 Türk Silahlı Kuvvetleri Örneği, unpublished doctorate dissertation, Harp Akademileri Komutanlığı Stratejik Araştırmalar Enstitüsü, İstanbul, 2011.

King, Anthony, *The Combat Soldier Infantry Tactics and Cohesion in the Twentieth and Twenty-First Centuries*, Oxford University Press, Oxford, 2013.

Krepinevich, Andrew and Watts, Barry, *The Last Warrior*, Basic Books, New York, 2015.

Kugler, Richard L., Operation Anaconda in Afghanistan: A Case Study of Adaptation in Battle, National Defense University, Washington, 2007.

Lieber, Keir A., "Grasping the Technological Peace the Offense-Defense Balance and International Security", *International Security*, 25:2, 2000, pp. 71–104.

McChrystal, Stanley, "It Takes a Network: The New Frontline of Modern Warfare", *Foreign Policy*, 2011. https://foreignpolicy.com/2011/02/21/it-takes-a-network/ accessed on 10 June 2020.

Mearsheimer, John J., *Conventional Deterrence*, Cornell University Press, Ithaca, 1983.

Miller, Adam, Turkey-PKK Conflict: Rising Violence in Northern Iraq, https:// acleddata.com/2022/02/03/turkey-pkk-conflict-rising-violence-in-northern-iraq/ accessed on 10 June 2022

Mueller, Karl P., "Examining the Air Campaign in Libya", in (eds.) Karl P. Mueller, *Precision and Purpose: Airpower in the Libyan Civil War*, RAND, Santa Monica, 2015, pp. 1–10.

Murray, Williamson, "Thinking About Revolutions in Military Affairs", J*oint Force Quarterly*, 16, 1997 pp. 69–76.

Ogarkov, Nikolai Vasilyevich, "The Defense of Socialism: Experience of History and the Present Day", *Red Star*, May 9, 1984 cited in Barry Watts, *The Evolution of Precision Strike*, Center for Strategic and Budgetary Assessments, Washington, 2013.

"Özel Harekatçı sayısı 40 bine çıkıyor", http://www.hurriyet.com.tr/gundem/ ozel-harekatci-sayisi-40-bine-cikiyor-40071325 accessed on 14 June 2020.

Özgen, Cenk, "Professionalization Tryouts in Turkish Armed Forces", *Trakya University Social Science Journal*, 13: 1, 2011, pp. 192–208.

Özgen, Cenk, "An Example of Conversion in the Security Field from 1980s to 2000s: Treatment of the Legal Arrangements Regarding Professional Military Service in Mainstream Journals", *Uludağ University, Social Sciences Journal*, 14:24, 2013, pp. 93–107.

Pamukoğlu, Osman, *Unutulanlar Dışında Yeni Bir Şey Yok*, İnkılap Kitapevi, İstanbul, 2004.

Parker, Geoffrey, "The Military Revolution 1560-1660-A Myth?", *Journal of Modern History*, 48, 1976, pp. 195–214.

Pirnie, Bruce R., et al., *Beyond Close Air Support Forging a New Air-Ground Partnership*, RAND, Santa Monica, 2005.

"PKK'nın değişen taktiklerine karşı yeni strateji", http://www.hurriyet.com.tr/ gundem/pkknin-degisen-taktiklerine-karsi-yeni-strateji-40094019 accessed on 02 October 2021.

Rumsfeld, Donald H., "Transforming the Military", *Foreign Affairs*, 2002, 81:3, pp. 20–32.

"Şemdinli Şafak Operasyonu" https://www.hurriyet.com.tr/gundem/semdinli-safak-operasyonu-21563693 accessed on 15 March 2020.

Sloan, Elinor, *Military Transformation and Modern Warfare A Reference Handbook*, Praeger, Westport, 2008.

Sullivan, Gordon R. and Dubik, James M., *Land Warfare in the 21st Century*, U.S. Army War College Strategic Studies Institute, Washington, 1993.

Sünnetçi, İbrahim, "İHA'lar ve Türkiye'nin İnsansız Havadan İstihbarat Çalışmaları", *Savunma ve Havacılık Dergisi*, 132, 2009, pp. 75–80.

"Türk Silahlı Kuvvetleri ve Genel Maksat Helikopterleri (2)", https://www. savunmasanayist.com/turk-silahli-kuvvetleri-ve-genel-maksat-helikopterleri-2/ accessed on 30 November 2021.

"Türkiye'nin milli silahlı hava aracı Bayraktar TB2'ler rekor kırdı",https://www. trthaber.com/foto-galeri/turkiyenin-milli-silahli-hava-araci-bayraktar-tb2ler-rekor-kirdi/40860/sayfa-5.html accessed on 26 November 2021.

"Türkiye'nin SİHA gücü ne kadar? Hangi modellerde kaç tane var? Çatışmalarda hangileri tercih ediliyor? İşte cevapları", https://www.indyturk.com/node/342926/ haber/t%C3%BCrkiyeninsi%CC%87hag%C3%BCc%C3%BC-ne-kadar-hangi-modellerde-ka%C3%A7-tane-var-%C3%A7at%C4%B1%C5%9Fmalarda accessed on 05 April 2021.

United States Special Operations Command Fact Book, USSOCOM Office of Communication, 2017.

Vickers, Michael G. and Martinage, Robert C., *The Revolution in War*, Center for Strategic and Budgetary Assessments, Washington, 2004.

Warner, Michael, "Reflections on Technology and Intelligence Systems", *Intelligence and National Security*, 27:1, 2012, pp. 133–153.

Watman, Kenneth H. and Raymer, Daniel P., *Airpower in US Light Combat Operations*, RAND, Santa Monica, 1994.

"100000 hours for gendarmerie", Twitter account of Gendarmerie General Command, accessed on 26 November 2021.

8 Military Innovation and Turkish Society

Toward a Post-Heroic Warfare?

Barış Ateş

Introduction

In developed Western countries, military technological innovations were associated with the post-Cold War era theory of casualty avoidance, or, as Luttwak proposed, post-heroic warfare. Luttwak argued that, as birth rates fell and the nuclear family emerged, the rate at which war casualties were accepted gradually declined. According to this theory, the significant decline over the 20th century in the average number of children per family in Western societies resulted in children being lost to war becoming less acceptable for families than it had been in the past.[1] In another article, Luttwak argued for decision-makers at the Pentagon to rarely recommend using ground forces, as they naturally have a high casualty risk due to their proximity to enemy fire. Therefore, the use of standoff weapon systems became more widespread.[2]

Combining the advantages of modern weapon systems with appropriate military planning makes carrying out armed, but almost bloodless, interventions possible. Of course, US goals should have been correspondingly modest and firmly adhered to the principles of partial, limited, and often slow results.[3] Bloodless wars, somewhat sarcastically stated and associated with new weapon systems, were a perfectly reasonable and acceptable solution for political and military leaders and society. In other words, advanced military technology and innovations were seen as saviors at a time when society did not want to lose their children in distant countries. Political leaders were spared the cost of increased casualties, and military leaders enjoyed the benefits of the new technology. Who would be bothered by a rapid conclusion to the Iraq and Afghanistan invasions, with very few casualties for military operations of that scale?

Based on post-heroic warfare theory, the necessary conditions for the post-heroic wars of the post-Cold War era to be conducted can be summarized as follows. On one hand, the pressures brought by societal changes, the decreased number of children, increased welfare, and the desire to live a comfortable life put pressure on governments. On the other hand, advanced weapon systems and the accompanying military innovations came into play to reduce this pressure. However, in contrast with Luttwak's theory, Peter D. Feaver and Charles Miller argued that society's "sense of casualties" only

DOI: 10.4324/9781003327127-9

becomes an issue when the usefulness of warfare is questioned.[4] In other words, they claimed that society would not be sensitive to casualties when war has a moral value. In fact, a similar approach is found in the discussions about heroism.

Scheipers stated that a hero can be defined as someone who risks their life for a cause. Ideally, this cause should have a political or moral value, or more abstractly, a "higher" value such as freedom from foreign domination. However, this was not the case for Western states. In other words, the states that had completed their development and solved their security problems did not expect a threat against their territorial integrity. Therefore, they could not produce any important or compelling political value to justify death on the battlefield. At this point, Scheipers argued that Western policymakers had justified military operations using political rhetoric. For example, humanitarian interventions in the 1990s were legitimized as measures to end genocide and protect human rights. The subsequent "war on terror" was waged in the name of civilization, freedom, and democracy; national self-determination also inspired recent Western operations in Libya. However, whether these universal values are as effective as nationalism at creating heroes is debatable. Universal values can appeal to all humanity, but these sacrifices are too weak to justify them and inspire heroism. In this case, Scheipers suggests that only the more particular values one associates with the idea of nation are able to produce heroes.[5]

The discussion outlined above tells the story of advanced Western democracies. However, how valid is this discussion in a geography like Türkiye's, which is between the West and the East and close to conflict zones? On one hand, Türkiye as a NATO member is a country in close relations with the West; it is pursuing EU membership – albeit unsuccessfully and unpromisingly – and has structured all its institutions, especially the military, according to Western standards. From this point of view, the above discussion could also be considered adaptable to Türkiye. But on the other hand, Türkiye does not have the security comfort that Western countries do. First of all, it has been dealing with a terrorist organization that aims to break its territorial integrity and internal peace for almost 40 years. It also borders regions of instability and conflict such as Syria and Iraq. The exaggerated threat assessments and maximalist demands of NATO-member Greece are another pressure point. Türkiye is one of the countries that has been key to the refugee problem occupying the world's agenda in recent years. In other words, Türkiye needs to ensure its own security in this case rather than fight for some universal values that are used to justify out-of-area military interventions. This situation explains why nationalism and patriotism have gained a strong place in Turkish society.

From the sociological point of view, the decrease in fertility rates, rise of the nuclear family, and increase in prosperity compared to the 1990s, and the new generation's awareness of the rest of the world both on the internet and in traditional ways are signs that Türkiye has entered a process similar to that of Western countries. In this case, has the casualty sensitivity apparent in the

West come to the fore in Türkiye as well? In other words, does Turkish society want to solve its problems and ensure its security through bloodless wars? In order to claim that the post-heroic war is also a phenomenon for Türkiye, a range of variables need to be examined.

First of all, the new military technology that has made a breakthrough in recent years needs to be taken into account, and the organizational and doctrinal changes brought about by this technology should be considered. Türkiye has a high perception of threat, with its ongoing fight against terrorism, and concern for protecting its own homeland. Therefore, fighting is unavoidable. But is this war occurring with the fewest possible casualties, or even being bloodless, something that is desirable? This bloodless option was not possible until a decade ago. Today, however, new military technology and the accompanying organizational and doctrinal changes have created an opportunity.[6] In this case, military innovation could be predicted to trigger the post-heroic war model. However, the army-society relationship and its differences with Western countries should be examined in detail, in addition to the advantages of military innovation movements and their results on the battlefield, and the threat perception of society and the state, in order to understand whether Türkiye faces issues such as post-heroic war or the desire to avoid casualties. This is exactly the purpose of this section.

The structure of this chapter is as follows. First, it compares the societal factors of Türkiye and Western countries. Additionally, the relationship between Turkish society and the military is analyzed separately due to some peculiar aspects different from Western democracies. Next, it explores military technologies that have emerged and their associated innovations to demonstrate the difference between the Türkiye of the 1990s and of today. Finally, it analyzes whether Türkiye has entered the post-heroic warfare era or not.

Societal Indicators

Sociologists have validated societal developments confirming Luttwak's post-heroic warfare theory. According to Moskos et al., changes in social and cultural values through modernity have caused a decrease in patriotism, a loss of belief in masculinity, and a rise of individualism in Western democracies. In addition, a materialistic lifestyle and the new information age have also contributed to this trend.[7] According to Manigart, individualism and hedonism have also become common values in Europe, with post-materialistic values having come to the fore. Duties to the nation and society have become meaningless, and the importance of traditional values has decreased. Soldiers are no longer motivated by the concept of patriotism. One of the most important results of this cultural change is that personal interests have been prioritized, and the desire for and belief in belonging to a group have begun to disappear. As a natural consequence of this, trust in institutions has decreased.[8]

Along with these developments, low birth rates, prolonged life expectancy, and immigration policies have brought about a significant change in being able to meet armies' personnel needs. For example, while the population between the ages of 15 and 29, which constitutes the age range the army requires, was eight million in Spain in 1999, estimates show that this number will decrease to four million by 2050. For NATO Europe, this number was 105 million in 2006 and will likely be 70 million in 2050.[9]

The situation in Türkiye should be noted to be different from the above countries. Türkiye does not have the same cultural values as the NATO countries with which it collaborates on defense matters. As seen in Inglehart and Welzel's study, many NATO countries attach more importance to individualism and quality of life than economic and physical security, and are more concerned with environmental or other humanitarian issues.[10] On the other hand, Türkiye is generally not included in the group of NATO countries with regard to cultural values such as self-expression, free decision-making, and individualism. Instead, it is located at a point between the East and West in accordance with its geography. Therefore, values that have become widespread in Western societies such as hedonism, resistance against authority, and an increased critical approach cannot be said to have started becoming popular in Türkiye in the full sense or with the same intensity. As a matter of fact, survey studies conducted in recent years have proven a similar result.[11]

For instance, Türkiye scores high on the power distance dimension [66], indicating its dependent and hierarchical characteristics. Turkish society is also collectivistic with a score of 37, signifying the sense of group belonging as important. Both indicators are much higher than other NATO countries.[12] Additionally, Türkiye is the leading country in Europe in terms of gender inequality, with 61.8% of society opposed to homosexuals taking part in military service, and 81.8% opposed to conscientious objection. The rate of those opposed to women in military service at 73.8% is also very high.[13] One recent study found 75.8% of society do not want to have homosexuals as their neighbors.[14]

Another justification for Luttwak's post-heroic warfare thesis involves improved living conditions alongside declining birth rates. A significant social change that took place in this period seems to require Türkiye to adopt post-heroic warfare, as birth rates have also declined in Türkiye. While the birth rate had been 3.07 in 1990[15] right after the Cold War, this rate dropped to 2.38 in 2001, and finally to 1.88 in 2019.[16] Another statistic that hints at adopting post-heroic warfare is average life expectancy. While the average life expectancy for the male population had been 65.6 years in 1990,[17] it increased to 78.6 in 2019.[18] In addition, while the per capita income had been $2,310 US in 1990, it increased to $12,630 US in 2014 then decreased to $9,690 US by 2020.[19] In other words, living conditions are seen to have improved on one hand while the number of children has reduced on the other. Therefore, casualty aversion may have developed in Turkish society. Ultimately, parents expect their children to receive a good education and be provided with all the opportunities that had not been available to previous generations. Therefore, the already decreasing number of children shows the aim of providing better opportunities to a smaller population. As such, when considering societal factors, Türkiye could be claimed to have both similarities and differences with Western countries. However, other variables should be reviewed before reaching any conclusion.

Public Support for the Military

When examined in terms of societal trust in their armies, a quite different situation emerges for Türkiye compared to other NATO countries. Confidence in the military in Türkiye increased from 91% in 1990 to 94% in 1997.[20] For the same period, while the rate of people with great confidence in the army was 61% in Türkiye, this rate was 33% in the USA, 17% in Canada, and 7% in Spain.[21] According to a survey conducted in 2011, 66% of the participants had full confidence in the military, and 18% had partial trust. In the same study, the rate of those with no trust in the army was only 15%. Of the participants, 13% were identified to "Partially agree" with the statement "Turks are a nation of soldiers", while 73% were seen to "fully agree". Moreover, 74% of the participants opposed the idea that compulsory military service should be abolished. Of the participants, 64% said that the army should consist of professional soldiers (20% partially agreed, plus 44% fully agreed).[22]

According to a study published in 2012, trust in the army was seen to have started declining as of 2008 and to have reached the level of 76%.[23] The interesting aspect of the issue is that, despite trust in civilian administrators and support for democracy having increased, trust in the army is still higher than in other institutions.[24] According to a survey conducted in 2013, confidence in the army was at 63.9%. Although this rate seems to be relatively low compared to previous years, it is high when compared to other institutions.[25] In that same study, the rate of those who said, "We should switch to a professional army" was 65.4%, while 44% of the participants were of the opinion that "Ending compulsory military service would harm the military-society bond".[26] Finally, according to a study conducted in 2017, trust in the military was at 85.1% and higher than for any other institution.[27] Turkish society can be said to have no clear stance on compulsory military service. The rate of those who oppose the abolition of compulsory military service, while at the same time want an army composed of professional soldiers being so high demonstrates society's double-edged approach.

Some historical reasons exist as to why confidence in the army is higher than for other institutions, even under the most adverse conditions. According to Tanel Demirel, the majority of events that have been accepted as successful in modern Turkish history were military in nature (e.g., the War of Independence, the Korean War, the Cyprus Peace Operation, and lastly, the war against PKK/KCK/YPG terrorism). This perception makes the army even more important as an institution for a society that has felt a backwardness in the face of the West for the last 300 years, and that has been unable to eliminate its inadequacy complexes in many areas.[28] According to the observation of a foreigner, many Turks think Westerners show prejudice toward them, and an inferiority complex emerges as a result. The military is accepted as the place where this pattern breaks, as it is instead seen and honored on equal terms with the West.[29] Moreover, thanks to compulsory military service, the fact that at least one person from each family has served in the military has enabled society to see the army as a part of itself.

In addition to the reasons mentioned above, some factors, whose roots reach much deeper, increase the value the military possesses in the eyes of society. This aspect can be explained by comparing Western and Turkish societies. The historical foundations of the armies in the West are not as strong as in Türkiye. For example, 18th-century officer cadres in Europe were designed for the needs of the aristocracy rather than the requirements of the military profession. Wealth, birth, and personal and political influence were the determining factors in the appointment and promotion of officers. Children and incompetent persons often achieved high military ranks. No one with professional knowledge was available. Therefore, apart from a few technical schools, no institution existed for transferring military knowledge, nor any system to put this knowledge into practice. The officers displayed aristocratic behaviors and belief styles rather than the behaviors required by the military profession.[30] In short, the military elite of the declining aristocracy in the West joined the army eager to find work and social status.[31] In this system, the uneducated noble officers who were in charge of the soldiers from war to war often failed to fulfill their duties.

The presence of officers from the noble class continued into the 20th century. Even in the second half of the 20th century, leaders from aristocratic backgrounds were found among American officers.[32] On the other hand, military service ceased to be a desirable profession due to the spread of anti-militarism after the heavy losses experienced in World War II. As a result, attracting qualified people to the military in countries possessing a professional military has become a problem in itself. In this case, the aims and motivations of protecting the homeland have been replaced by financial means.[33]

No such process has occurred in Türkiye. The officer class has always consisted of people from a broad stratum, including in the Ottoman Empire period.[34] Moreover, officers are found to have formed a unity of destiny with society. Before the Ottoman Empire lost a large part of its territory and eventually turned into a young republic established in Anatolia, many displaced people were found within the empire's territory. A special bond was established between these groups and the army due to the lack of solid civilian institutions. In the Balkans especially, the only official authority that citizens could look to as a representative from the state was the regional commander and his soldiers. While they were trying to protect the Muslim Turks against rebels, they also undertook education, health, construction, and afforestation efforts in the towns and villages they were responsible for. When necessary, they built roads, bridges, mosques, and schools and also took on other public services. Therefore, very close ties and a sense of solidarity were established between the people and the soldiers.[35] This bond that developed between society and the soldier can safely be said to continue to this day.

The Turkish military has no aristocratic tradition. Therefore, mentioning any elite tradition based on ancestry and nobility is impossible; Ottoman pashas' aristocracy did not form like the old Western nobles who would go on to find a new job in the army. As a result, while an emerging new class with the weakening of feudalism in Western societies had been able to design the state and the military according to its own social understanding, this process involved the state and military designing Turkish society. First a professional military tradition and state centralism had formed, from which no bourgeoisie were able to emerge. Moreover, the great transformation experienced in the 1920s was realized thanks to the army. At that time, no other social class or power apart from the military existed that could establish the Republic in Türkiye.[36] The fact that the military and society are so intertwined explains why the army is more trusted than other institutions, even in the most challenging conditions. Moreover, because technological support had remained fragile in military operations until the last few years, personnel-based strategies had to be followed. These significant differences between Western armies and the Turkish army explain the considerable public support.

Military Technology and Associated Innovations

Fifteen years ago, Türkiye lacked sufficient advanced technological weapon systems such as uncrewed aerial vehicles (UAVs), national warships (MİLGEM), advanced electronic jamming systems, and smart ammunition. At that time, the main component of its search-find-fix-destroy operations against terrorist elements was personnel -based. In particular, the commando units actively roamed the terrain, maintained control, and tried to make contact with, and neutralize terrorist groups. When any intelligence was received from images obtained through limited surveillance means, the airborne operation or unmounted troops would sometimes conduct a search operation that lasted for days. In this type of operation, the main determinants were the physical conditions; terrain analysis using classical methods and reconnaissance, surveillance, and concealment; and the shooting skills of individual soldiers. To summarize, the units that knew how to move in the field, that had good command of their weapons, and that were able to perform properly their course of action (i.e., basic military skills), such as surveillance, concealment, and reconnaissance activities would be successful. However, personnel-based tactics were also reflected in the casualty rates during this period. Thousands of soldiers, police, and security guards were killed or injured; despite this, however, they successfully saved the country's territorial integrity. In fact, the Turkish Army's fight against terrorism has been among the few successful counter-terrorism operations in the world.[37]

During this period, the time required to locate and destroy a terrorist group could take days and sometimes weeks. As personally experienced by the author, troops had to march hundreds of kilometers in some operations. These harsh conditions helped raise qualified troops but at the same time spread the struggle over time; the entry of terrorists into the country could not be prevented due to the lack of technology and the harsh terrain. On the border of Iraq and Iran, the steep and snowy mountains above 3,000 meters could not be adequately observed. Naturally, society knew of these difficulties and problem areas during this period when the compulsory soldiers were actively involved in these operations. Everyone was suggesting solutions to how to win the war against terrorism, but some of these proposed solutions had the potential to threaten internal peace. However, the issue upon which everyone agreed was the need for high technology weapon systems. Political and military leaders as well as society in general wanted the military to have high-tech weapons as part of the solution to the problem. In this environment, weapon systems mainly produced by Türkiye's domestic resources played a role in changing the balance.

While the soldiers at temporary or permanent operation bases and border outposts had in the past taken precautions against infiltration attempts by relying on their creativity, together with a limited number of surveillance tools, sensors and thermal cameras today are able to detect the footsteps of even a rabbit approaching the base. Moreover, terrorist groups can be placed under fire with remote-controlled weapon systems. Most importantly, thanks to the development of communication technologies, a whole process from the first moment of fire to the last step of casualty evacuation can be carried out and coordinated easily and quickly. Similarly, only mine detectors and visual inspections were available in the past for finding and neutralizing improvised explosive devices (IEDs). Today, the Turkish military probably has the best and most experienced IED teams equipped with high-tech products, thanks to domestic technological improvements. In light of these experiences, the Turkish Armed Forces (TAF) is seen to have increasingly focused on technology-intensive operations and achieved significant successes, especially with the development of a domestic defense industry.

The need for troops to search for days for terrorists in the field no longer exists. Systems that were just a dream 30 years ago are now deployed and actively used even in the smallest military units. Armed and unarmed UAVs, extremely light and functional thermal systems, remote-controlled weapon systems, technologies that provide instant geographical analysis, and light but practical clothing and equipment have provided significant advantages to the troops. In the past, a search-find-destroy operation carried out by a commando unit in a week can now be completed in a few hours by drones. When any terrorist element is detected, advanced target designation systems and fire support weapons come into play. This situation naturally reduces the number of casualties and increases military effectiveness. The commando brigades established during this period are entirely composed of professional soldiers and have carried out highly successful operations in areas such as Iraq and Syria.

All these advanced systems put into the service of the army have of course been appreciated by society. From new weapon systems to soldiers' uniforms, almost every new military product has been discussed daily in the press and social media. Images of UAVs or MİLGEM ships are a source of pride. Of course, having a strong army is desirable. In one current study, the desire to have a strong military was chosen as the second most important priority by 38.5% of society. A high rate of economic growth came first.[38] This amount of social support helps military innovation projects be carried out more easily.

At the other end of the innovations brought by military technology is the increased rate of professionalization in the army. Contrary to developed Western democracies, the Turkish military still applies the conscription model. However, this is actually a mixed model. In other words, while the military does have conscription on one hand, it also has a professionalization rate approaching almost 50% on the other hand. Moreover, compulsory soldiers have not been deployed in operations or at-risk areas for the last ten years.[39] In this way, the bond between the army and society is preserved while simultaneously increasing military efficiency. In other words, political and military decision-makers can act more comfortably with regard to military requirements. However, unlike Luttwak's theory, avoiding casualties in these operations is not the first priority. Furthermore, Turkish military planners are observed not to hesitate in using ground troops. In other words, no matter how much the technology develops, the principle of having boots on the ground continues, which has been reflected in the operations in both Syria and northern Iraq.

Toward a Turkish Post-Heroic Warfare?

During the counter-terrorist operations that began in the 1980s, Türkiye has lost thousands of security personnel and civilians. People from almost every part of society have been impacted both physically and psychologically. Of course, by feeling the threat so close, Turkish society expected effectiveness in anti-terrorist operations. This expectation does not mean that society does not react to the loss of soldiers or civilians. However, to say that the price is at an acceptable level for a society that feels the threat so closely would not be an exaggeration. Indeed, ten years after Luttwak introduced his post-heroic warfare theory, conscription applications were still peaking in Türkiye. Conscientious objection was limited to an extremely marginal group. Soldiers were sent to their new units with drums and *zurna*s (shrill wind instruments). Varoğlu and Bıçaksız demonstrated with historical examples and empirical data that Turkish society has a high degree of acceptance of risk and loss. Even in the years when casualties hit their peak, no decrease was seen in applications for either compulsory military service or professional military positions. In fact, the recruitment pool was quite large.[40]

Between 1994 and 2002, an average of 20 applicants was generally found for each job position as a non-commissioned officer, and applications for the officer category were traditionally high, with the selection rate being 50:1 or higher.[41] As for conscripted soldiers, military service is seen as a rite of passage and proof of masculinity in Türkiye. Soldiers' willingness to serve in operationally risky areas should be studied against this background.[42]

Additionally, a part of TAF culture involves making sure neither martyrs nor their families are forgotten. Specific policies and programs are in place for the latter, including financial and non-material support. Units and schools such as military academies have memorials and cemeteries to commemorate veterans and martyrs. Another way of showing society's respect is to name parks, streets, and schools after martyrs.

Recent applications to become a professional soldier also show that society still maintains a martial spirit. Thousands of young people have applied to become officers or non-commissioned officers. Of course, this has something to do with the deteriorating economic balance in recent years. For example, while 43 candidates were still found for each officer position in 2016, 35 applications were received for each non-commissioned officer position.[43] The figures here also coincide with the data from Varoğlu and Bıçaksız's study. Moreover, these years coincide with the period when the most intense clashes were experienced in domestic and international operations. In other words, whether or not Türkiye has entered the era of post-heroic warfare is still too early to tell, although the social conditions mentioned by Luttwak have been realized in Türkiye. Societal support should be noted to continue on this point, with no casualty-aversion having developed yet. Of course, society reacts very strongly to news of every loss, but at the same time, it acknowledges the necessity of operations. As a matter of fact, two different surveys showed the public support for Operation Olive Branch, a cross-border military operation conducted by the Turkish Armed Forces and Syrian National Army in the Afrin District of northwest Syria to have been over 80%. While this rate had been 85%[44] in the first survey conducted in January 2018, it was shown to be 89%[45] in another survey conducted at the same time.

The most important reason for this involves the threat perception and societal concerns about the future of the country. According to a recent study, the rate of those who are worried that the country will be dragged into a war is 84.9%, and 85.6% are concerned about a terrorist attack. These respective rates are 36% and 40.1% in Greece, 74.6% and 67.8% in the USA, and 39.1% and 57.8% in Germany.[46] The high percentages in Türkiye also substantiate why society considers heroism to be so valuable. Rather than intervening in a problem that arises in a distant part of the world, the concern relates to their own survival. Therefore, to sacrifice one's life for the sake of homeland and nation is not so surprising in Turkish society. This situation also reveals the difference regarding casualty sensitivity and post-heroic warfare that exists between the West and Türkiye. In other words, Feaver and Miller's thesis that sensitivity to casualties becomes a problem when the purpose of war is questioned is confirmed in the case of Türkiye. When the aim of war is the security of the country, the values of nationalism predominate, and the concept of heroism, being fed by this nationalism, gets reinforced.

According to Bozdemir, the common feature of previous and current Turkish states is the importance attached to military service and the army.[47] In parallel with this, military service and warriorship have always been glorified values. Society always speaks of its soldiers as heroes. In fact, civilians who make outstanding sacrifices in other professions are also called heroes. For example, a firefighter injured while responding to forest fires is also considered a hero. The news from conflict zones also emphasizes heroism. The word "hero" is always included in statements and social media messages published by the Ministry of National Defense. In short, society and the state bureaucracy, feeling themselves under imminent threat, cherish heroism. Although casualties are questioned, tolerance is generally higher than in Western countries. Being far removed from conflict zones, citizens of welfare states in the West only know about the issue if a relative is in a conflict zone or if they are consciously following the situation of their soldiers. On the other hand, Türkiye has been dealing with threats from neighboring countries for the last 40 years, as well as fighting against a terrorist organization that directly targets its own citizens and carries out all kinds of violence without hesitation. This situation results in a high tolerance to casualties.

The Other Side of the Coin

I stated above that the post-heroic warfare has not developed in Türkiye and that the degree of acceptance of casualties is higher than that of Western societies. However, being cautious at this stage and allowing room for a potential increase in casualty aversion would be better. This is because, despite no empirical study having been conducted yet on measuring casualty tolerance in society, some signs that a careful observer would notice indicate that the degree of casualty acceptance may yet decrease in the near future and that casualty sensitivity will increase in society and also among military and political leaders accordingly. The first indicator in this regard is the question that comes to the fore both in the written and visual media, especially on social media after any news of a casualty: "Don't we have UAVs? Why didn't we solve the problem by sending a few drones instead of putting our soldiers at risk?" This type of question should now be seen as a natural reflex of a society that realizes the possibilities and capabilities of advanced technology and its resulting military effectiveness. In other words, society wants to see the results of the new weapon systems. However, probably the most important reason why this request is not expressed out loud is related to the size of the threat.

Another important issue that should be taken into consideration is the unity in society in regard to military service, especially in wartime. According to Moskos, casualty acceptance is the most crucial variable in the relationships between the armed forces and society. He argued that the focus should be on the social composition of those who join the military and those who do not. He emphasizes this point in particular: that society will accept high casualty rates only if the leaders are self-sacrificing. For example, in both world wars, the British "noble" class suffered more casualties than the "working" class.[48] During World War II, 223 of 675 German Generals (i.e., 33%) died in battle. The Germans determined the total number of the nobility class during World War II to have been 8,284, of whom 4,690 had died in battle.[49]

Despite these positive examples from the past, the situation today is quite different. For instance, children of only eight of the 535 members of the US Senate and Congress were serving in Iraq in 2004.[50] As a result, Moskos opined that national security and interests are ensured to be taken much more seriously when the children of the country's elite families take their military service seriously.[51]

Similar examples can be found in the history of Türkiye. The well-educated youth of the Ottoman Empire almost vanished during World War I and the subsequent Independence War. This, obviously, was one of the important factors in gaining the victory. However, today's militaries usually deal with counter-terrorism operations and peace enforcement issues involving low- and medium-intensity conflicts. In the Turkish case, the rate of casualties was higher in the early 1990s than today, but now news of the martyrdom of a single soldier draws a great reaction from society and leads to the entire state and army mechanism being questioned. This approach can also be seen in the news that has started to appear in media organs in Türkiye in recent years. Seeing the children of the upper social classes also participate in the fight against terrorism and pay their own price is desirable. However, the belief that children of families belonging to upper social classes in Türkiye are the least affected by compulsory military service and its consequences seems to have settled in the collective memory.[52] These reactions seem to have been influential in adopting the semi-professional army model. Conscripts no longer serve in risky areas or counter-terrorism operations. All these tasks are carried out by professional staff. This should be accepted as a sign that society's tolerance for casualties has decreased.

Another additional factor involves concerns about paid exemption from military service, which is the right to be exempted from military service in return for a certain fee. The first paid exemption from military service practice in Türkiye was implemented in 1987. After that, it was reenacted in 1992, 1999, 2011, and 2014. The paid exemption from military service arrangement made in 2018 was the last temporary one. In 2018, 600,000 people applied for paid exemption from military service, while this number had only been 18,433 in 1987. As of 2019, conscripted military service was reduced to six months and paid exemption from military service became permanent. This means that those who have a good financial standing and can pay the fee are exempted from the six months of military service.

These paid exemption from military service practices have generally led to the consolidation of the distinction between rich and poor in society. Therefore, those who completed conscription because they did not have money to pay for exemption, as well as their families, can perceive the practice of paid exemption from military service as a response for avoiding becoming a casualty as well as being a non-heroic situation. The negative approach created by the paid exemption from military service practices in the eyes of society is closely related to the debates initiated by the fact that the children of the leading figures in the country generally serve far away from operation areas. Additionally, the increasing awareness of the new generation about the rest of the world paves the way for comparing themselves to their peers. No empirical data currently exists for measuring this situation. However, the reactions shown at the funerals of martyrs, the way news is presented in printed and visual media, and debates on social media should be considered as preliminary signs that the tolerance for casualties is gradually decreasing. Of course, heroic warriors are still considered valuable and respected members of society in Türkiye, but society wants to see more of the benefits of military technology. The argument can be made that as soon as their security concerns (threat perceptions) begin to wane, the level of casualty avoidance will increase and turn into a challenge for political and military leaders. The prediction can be made that a country that has turned its face to the West, become a member of almost all Western international organizations, and has a military organization and doctrine resembling the West will experience similar developments, albeit at a later time. Therefore, the military needs to find rational solutions to the problems it may encounter by taking into account the experiences of its allies.

Conclusion

This chapter has examined whether new military technological and social developments are the factors that undermine heroism and reduce casualty tolerance in Türkiye. In this context, both similarities and differences are seen with Western countries that have experienced post-heroic warfare. For example, nuclear families with few children now exist in Türkiye, and the level of welfare has improved considerably compared to the Cold War period. In addition, the Turkish Army now has the means and capabilities to destroy an enemy from afar, and these innovations have changed the balance on the battlefield.

However, one significant difference is seen to exist between Türkiye and the West with regard to threat perception. Another important difference is the place the army has in the eyes of society as well as society's trust in the military. Even in the most challenging of times, the confidence in the military has remained high. As a natural consequence of this trust, the military should be strong and effective because countering the threat to the integrity of the country is only possible with a strong military. Therefore, tolerance toward casualties remains high. In addition, heroism is considered a noble value that is always to be exalted. In other words, the tradition of the heroic warrior is maintained both in the army and in society. Therefore, although military planners consider minimum casualties in their operational planning, this does not prevent putting a high number of ground troops on the battlefield.

When comparing these variables, it could be said that military innovation does not automatically develop into post-heroic warfare, but it does pave the way for it. Therefore, with the elimination of threats and the advent of military innovations, signs are also found to indicate that casualty avoidance, which has yet to change military priorities, could be emphasized more strongly. Of course, due to Türkiye's solid cultural structure, post-heroic warfare might not occur in Türkiye as it has in the West; however, the desire for casualty avoidance might increasingly become an element of pressure for political and military leaders.

Notes

1 Edward N. Luttwak, "Toward Post-Heroic Warfare", *Foreign Affairs*, 74:3, May/June 1995, p. 115.
2 Edward N. Luttwak, "A Post-Heroic Military Policy: The New Season of Bellicosity", *Foreign Affairs*, 75:4, July/August 1996, p. 37.
3 Luttwak, "Toward Post-Heroic Warfare", p. 114–115.
4 Peter D. Feaver and Charles Miller, "Provocations on Policymakers, Casualty Aversion and Post-Heroic Warfare", in (ed.) Sibylle Scheipers, *Heroism and the Changing Character of War*, Palgrave Macmillan, Hampshire, 2014, pp. 145–161.
5 Sibylle Scheipers, "Introduction: Toward Post-Heroic Warfare?", *Heroism and the Changing Character of War*, in (ed.) S. Scheipers, Palgrave Macmillan, Hampshire, 2014, p. 1–4.
6 See Chapters 6 and 7 in this book.
7 Charles Moskos, "Military Systems in the 21st Century: Changes and Continuities", in (eds.), Timothy Edmunds and Marjan Malešič, *Defence Transformation in Europe: Evolving Military Roles*, IOS Press, Amsterdam, 2005, p. 23.
8 Philippe Manigart, "Restructuring of the Armed Forces", in (ed.) G. Caforio, *Handbook of The Sociology of The Military*, Springer, New York, 2006, pp. 326–327.
9 Cindy Williams, Curtis Gilroy, "The Transformation of Personnel Policies", *Defence Studies*, 6:1, 2006, p. 104.
10 Ronald Inglehart, Christian Welzel, *Modernization, Cultural Change and Democracy: The Human Development Sequence*, Cambridge University Press, Cambridge, 2005, pp. 137–141.
11 Yılmaz Esmer, *Değerler Atlası*, 2012, https://bau.edu.tr/haber/1725-turkiye-degerler-atlasi-2012-yayinlandi, accessed on 17 Aug 2021.
12 Hofstede Insights, https://www.hofstede-insights.com/country-comparison/turkey/, accessed on 26 November 2021.
13 Yaprak Gürsoy, Zeki Sarıgil, Türkiye'de Silahlı Kuvvetler ve Toplum: Ampirik Yaklaşım, http://bilgi.edu.tr/tr/haberler-ve-etkinlikler/haber/536/turkiyede-silahl-kuvvetler-ve-toplum-anket-sonuclar/4.4.2012, accessed on 25 August 2021.
14 World Values Survey(2017–2020), www.worldvaluessurvey.org/WVSDocumentatin WV7.jsp?COUNTRY=3460, accessed on 21 July 2021. Q22 – On this list are various groups of people. Could you please mention any that you would not like to have as neighbors? Homosexuals: %75.8
15 https://data.oecd.org/pop/fertility-rates.htm, accessed on 24 July 2021.
16 https://data.tuik.gov.tr/Bulten/Index?p=Dogum-Istatistikleri-2019-33706, accessed on 24 July 2021.
17 http://www.healthdata.org/turkey, accessed on 24 July 2021.
18 https://data.tuik.gov.tr/Bulten/Index?p=Hayat-Tablolari-2017-2019-33711, accessed on 26 July 2021.
19 https://data.worldbank.org/country/turkey?locale=tr, accessed on 27 July 2021.
20 Yılmaz Esmer, *Devrim, Evrim, Statüko: Türkiye'de Sosyal, Siyasal, Ekonomik Değerler*, TESEV Yayınları, İstanbul, 1999, p. 42.
21 Zeki Sarıgil, "Deconstructing the Turkish Military's Popularity", *Armed Forces & Society*, 35:4, July 2009, p. 710.

22 Gürsoy and Sarıgil, Türkiye'de Silahlı Kuvvetler ve Toplum: Ampirik Yaklaşım, http://bilgi.edu.tr/tr/haberler-ve-etkinlikler/haber/536/turkiyede-silahl-kuvvetler-ve-toplum-anket-sonuclar/4.4.2012, accessed on 25 Aug 2021.

23 Esmer, Türkiye Değerler Atlası 2012.

24 Sarıgil, "Deconstructing the Turkish Military's Popularity", p. 711.

25 Salih Akyürek, Mehmet Ali Yılmaz, *Türk Silahlı Kuvvetlerine Toplumsal Bakış*, Bilgesam, Ankara, 2013, p. 10.

26 Akyürek and Yılmaz, *Türk Silahlı Kuvvetlerine Toplumsal Bakış*, p. 45.

27 World Values Survey (2017–2020), www.worldvaluessurvey.org/WVSDocumentationWV7.jsp?COUNTRY=3460, accessed on 21 July 2021. Q65 – I am going to name a number of organizations. For each one, could you tell me how much confidence you have in them: is it a great deal of confidence, quite a lot of confidence, not very much confidence or none at all? The armed forces: great deal 46.6% and quite a lot 38.5%.

28 Tanel Demirel, "Türk Silahlı Kuvvetleri'nin Toplumsal Meşruiyeti Üzerine", in (eds.) Ahmet İnsel, Ali Bayramoğlu, *Bir Zümre, Bir Parti: Türkiye'de Ordu*, Birikim Yayınları, 4thed., İstanbul, 2009, p. 358.

29 "The Military and Turkish society", *The Adelphi Papers*, 41:337, 2001, p. 13.

30 Samuel P. Huntington, *The Soldier and the State: The Theory and Politics of Civil-Military Relations*, Vintage Books, New York, 1957, p. 28.

31 Mevlüt Bozdemir, *Türk Ordusunun Tarihsel Kaynakları*, Ankara Üniversitesi Siyasal Bilgiler Fakültesi Yayınları, Ankara, 1982, p. 180.

32 Morris Janowitz, *The Professional Soldier: A Social and Political Portrait*, Free Press, New York, 1964, p. 160.

33 M.Ali Birand, *Emret Komutanım*, Milliyet Yayınları, 8th ed, İstanbul, 1986, p. 26.

34 Ahmet Taner Kışlalı, "Türk Ordusunun Toplumsal Kökeni Üzerinde Bir Araştırma", *Ankara Üniversitesi SBF Dergisi*, 29:2, 1974, p. 89–105.

35 Mesut Uyar, Edward J. Erickson, *A Military History of the Ottomans: From Osman to Atatürk*, ABC-CLIO, California, 2009, pp. 214–216.

36 Bozdemir, *Türk Ordusunun Tarihsel Kaynakları*, pp. 168–176.

37 Christopher Paul, *Counterinsurgency Scorecard: Afghanistan in Early 2011 Relative to the Insurgencies of the Past 30 Years*, RAND, California, 2011.

38 World Values Survey (2017–2020), www.worldvaluessurvey.org/WVSDocumentationWV7.jsp?COUNTRY=3460, accessed on 21 July 2021. Q152 – People sometimes talk about what the aims of this country should be for the next ten years. On this card are listed some of the goals which different people would give top priority. Would you please say which one of these you consider the most important? Making sure this country has strong defense forces: 38.5

39 Metehan Demir, Çatışma bölgelerine profesyoneller gidecek, 21.11.2012, https://www.hurriyet.com.tr/gundem/catisma-bolgelerine-profesyoneller-gidecek-21976926, accessed on 29 July 2021.

40 A. Kadir Varoğlu and Adnan Bıçaksız, "Volunteering for Risk: The Culture of the Turkish Armed Forces", *Armed Forces & Society*, 31:4, 2005, pp. 583–598.

41 Varoğlu and Bıçaksız, "Volunteering for Risk", p. 592.

42 Varoğlu and Bıçaksız, "Volunteering for Risk", p. 594.

43 TSK'ya rekor başvuru, https://www.dunya.com/gundem/tskya-rekor-basvuru-haberi-340533, 5 Aralık 2016, accessed on 21 July 2021.

44 MAK Danışmanlık: Halkın yüzde 85'i Afrin operasyonunu destekliyor, 29 January 2018. https://www.politikyol.com/mak-danismanlik-halkin-yuzde-85-i-afrin-operasyonunu-destekliyor/, accessed on 21 July 2021.

45 https://www.milliyet.com.tr/siyaset/unlu-arastirmaci-adil-gur-zeytin-dali-harekatina-destek-yuzde-90-2608608, 12.02.2018, accessed on 21 July 2021.

46 World Values Survey (2017–2020), www.worldvaluessurvey.org/WVSDocumentation WV7.jsp?COUNTRY=3460 accessed on 21 July 2021. Q146 To what degree are you worried about the following situations? A war involving my country: Very much 39.2 A great deal 45.7. Q147 – To what degree are you worried about the following situations? A terrorist attack: Very much 42.7, A great deal 42.9.
47 Bozdemir, *Türk Ordusunun Tarihsel Kaynakları*, p. 3.
48 Moskos, "Military Systems in the 21st Century: Changes and Continuities", p. 23.
49 Richard A. Gabriel, Paul L. Savage, *Crisis in Command: Mismanagement in the Army*, Hill and Wang, New York, 1978, p. 36.
50 Moskos, "Military Systems in the 21st Century: Changes and Continuities", p. 24.
51 Charles C. Moskos, Lawrence Korb, "Two Views: Time to Bring Back the Draft?", *The American Enterprise*, 12:8, December 2001, p. 17.
52 Akyürek and Yılmaz, *Türk Silahlı Kuvvetlerine Toplumsal Bakış*, p. 18.

References

Akyürek, Salih and Yılmaz, Mehmet Ali, *Türk Silahlı Kuvvetlerine Toplumsal Bakış*, Bilgesam Rapor No:56, Ankara, 2013.
Birand, M. Ali, *Emret Komutanım*, 8th ed, Milliyet Yayınları, İstanbul, 1986.
Bozdemir, Mevlüt, *Türk Ordusunun Tarihsel Kaynakları*, Ankara Üniversitesi Siyasal Bilgiler Fakültesi Yayınları, Ankara, 1982.
Demir, Metehan, "Çatışma bölgelerine profesyoneller gidecek", *Hürriyet*, 21.11.2012, https://www.hurriyet.com.tr/gundem/catisma-bolgelerine-profesyoneller-gidecek-21976926, accessed on 29 July 2021.
Demirel, Tanel "Türk Silahlı Kuvvetleri'nin Toplumsal Meşruiyeti Üzerine", in (eds.) Ahmet İnsel and Ali Bayramoğlu, *Bir Zümre, Bir Parti: Türkiye'de Ordu*, 4thed., Birikim Yay., İstanbul, 2009, pp. 345–381.
Doğum istatistikleri, https://data.tuik.gov.tr/Bulten/Index?p=Dogum-Istatistikleri-2019-33706, accessed on 24 July 2021.
Esmer, Yılmaz, *Devrim, Evrim, Statüko: Türkiye'de Sosyal, Siyasal, Ekonomik Değerler*, TESEV Yayınları, İstanbul, 1999.
Esmer, Yılmaz, *Değerler Atlası*, 2012, https://bau.edu.tr/haber/1725-turkiye-degerler-atlasi-2012-yayinlandi, accessed on 17 Aug 2021.
Feaver, Peter D. and Miller, Charles, "Provocations on Policymakers, Casualty Aversion and Post-Heroic Warfare", in (ed.) Sibylle Scheipers, *Heroism and the Changing Character of War*, Palgrave Macmillan, Hampshire, 2014, pp. 145–161.
Fertility rates, https://data.oecd.org/pop/fertility-rates.htm, accessed on 24 July 2021.
Gabriel, Richard A. and Savage, Paul L., *Crisis in Command: Mismanagement in the Army*, Hill and Wang, New York, 1978.
Gürsoy, Yaprak and Sarıgil, Zeki, *Türkiye'de Silahlı Kuvvetler ve Toplum: Ampirik Yaklaşım*, http://bilgi.edu.tr/tr/haberler-ve-etkinlikler/haber/536/turkiyede-silahl-kuvvetler-ve-toplum-anket-sonuclar/4.4.2012, accessed on 25 Aug 2021.
Hayat Tabloları, https://data.tuik.gov.tr/Bulten/Index?p=Hayat-Tablolari-2017-2019-33711, accessed on 26 July 2021.
Hofstede Insights, https://www.hofstede-insights.com/country-comparison/turkey/, accessed on 26 November 2021.
Huntington, Samuel P., *The Soldier and the State: The Theory and Politics of Civil-Military Relations*, Vintage Books, New York, 1957.
Inglehart, Ronald and Welzel, Christian, *Modernization, Cultural Change and Democracy: The Human Development Sequence*, Cambridge University Press, Cambridge, 2005.

Janowitz, Morris, *The Professional Soldier: A Social and Political Portrait*, Free Press, New York, 1964.

Kışlalı, Ahmet Taner, "Türk Ordusunun Toplumsal Kökeni Üzerinde Bir Araştırma", *Ankara Üniversitesi SBF Dergisi*, 29:2, 1974, pp. 89–105.

Luttwak, Edward N., "Toward Post-Heroic Warfare", *Foreign Affairs*, 74:3, May/June 1995, pp. 109–122.

Luttwak, Edward N., "A Post-Heroic Military Policy: The New Season of Bellicosity", *Foreign Affairs*, 75:4, July/August 1996, pp. 33–44.

MAK Danışmanlık: Halkın yüzde 85'i Afrin operasyonunu destekliyor, 29 January 2018.

Manigart, Philippe, "Restructuring of the Armed Forces", in (ed.) G. Caforio, *Handbook of The Sociology of The Military*, Springer, New York, 2006, pp. 326–343.

Moskos, Charles C. and Korb, Lawrence, "Two Views: Time to Bring Back the Draft?", *The American Enterprise*, 12: 8, December 2001.

Moskos, Charles "Military Systems in the 21st Century: Changes and Continuities", in (eds.), Timothy Edmunds and Marjan Malešič, *Defence Transformation in Europe: Evolving Military Roles*, IOS Press, Amsterdam, 2005, pp. 19–26.

Paul, Christopher, *Counterinsurgency Scorecard: Afghanistan in Early 2011 Relative to the Insurgencies of the Past 30 Years*, RAND, Santa-Monica, 2011.

Sarıgil, Zeki, "Deconstructing the Turkish Military's Popularity", *Armed Forces & Society*, 35:4, July 2009, pp. 709–727.

Scheipers, Sibylle, "Introduction: Toward Post-Heroic Warfare?", in (ed.) Sibylle Scheipers, *Heroism and the Changing Character of War*, Palgrave Macmillan, Hampshire, 2014, pp. 1–18.

"The military and Turkish society", *The Adelphi Papers*, 41:337, 2001, 9–20.

TSK'ya rekor başvuru, https://www.dunya.com/gundem/tskya-rekor-basvuru-haberi-340533, 5 Aralık 2016, accessed on 21 July 2021.

Turkey, https://data.worldbank.org/country/turkey?locale=tr, accessed on 27 July 2021.

Turkey health data, http://www.healthdata.org/turkey, accessed on 24 July 2021.

Turkey, *World Values Survey (2017–2020)*, accessed on 21 July 2021. www.worldvaluessurvey.org/WVSDocumentationWV7.jsp?COUNTRY=3460.

Uyar, Mesut and Erickson, Edward J., *A Military History of the Ottomans: From Osman to Atatürk*, Praeger Security International, ABC-CLIO, California, 2009.

Varoğlu, A. Kadir and Bıçaksız, Adnan, "Volunteering for Risk: The Culture of the Turkish Armed Forces", *Armed Forces & Society*, 31:4, Summer 2005, pp. 583–598.

Williams, Cindy and Gilroy, Curtis, "The Transformation of Personnel Policies", *Defence Studies*, 6:1, 2006, pp. 97–121.

https://www.politikyol.com/mak-danismanlik-halkin-yuzde-85i-afrin-operasyonunu-destekliyor/, accessed on 21 July 2021.

https://www.milliyet.com.tr/siyaset/unlu-arastirmaci-adil-gur-zeytin-dali-harekatina-destek-yuzde-90-2608608, 12.02.2018, accessed on 21 July 2021.

9 Quick-Impact Approach in Information Warfare

Turkish Experience

Dağhan Yet

Introduction

In the new millennium, violence is slowly creeping away out of ordinary people's lives. It has decreased for thousands of years[1] as humans have transformed their societies from humble hunter-gatherers to that of the *homo deus*. As a result, modern war has become much more entangled with non-kinetic means over the last decades, as people do not want to experience violence, even in war. As Valery Gerasimov famously stated in his 2013 article, non-kinetic means dwarf kinetic means four to one.[2] The author lacks the expertise to claim the ratio that the Russian general suggested, but it is certainly possible that the military change is shifting towards a non-kinetic form in our contemporary era.

Gerasimov underlined the importance of these *ingenious ways*, with which Clausewitz would disagree regarding their *game-winning* aspects[3], but Gerasimov also emphasized that violence still plays the most critical role in contemporary conflicts.[4] As he goes on to argue, however, this critical role of conflict is relatively limited and should be exercised only on the spot and with surgical precision.

As Azar Gat claimed, violence, especially kinetic violence, is mostly not present in the lives of the developed world, and the world has become a more peaceful planet than ever before in its history;[5] everyone seeks a life devoid of violence. Nearly no-one wants a hero in their family anymore. Parents do not want their sons or daughters returning home in coffins; heroic warfare is dead.[6] Modern societies want to win wars without getting their hands dirty, or at least getting as minimally dirty as possible.

In the author's understanding, influence operations, including information and psychological operations, as well as other non-kinetic means such as diplomacy and politics have gained significant importance in modern warfare due to the decreasing levels of violence in our everyday lives. Everyone dreams of the victory that Sun Zi defined as the "ultimate excellence".[7] This tendency results in a misconception that playing the long game is the only way to win non-kinetic wars, as they do not require any form of violence. Contrarily, some conflicts need a quick-impact approach (QIA) in influence operations, which will be described in this chapter.

DOI: 10.4324/9781003327127-10

QIA is about the pacing of these influence operations. Any country or organization may choose to gain long- or short-term goals from these operations. As pointed out in the previous paragraph, the tendency exists to assume that all information and psychological operations that seek to destroy a nation from the inside should be long-term; nevertheless, as contrasting examples will show, these operations or tactics can also be waged for short-term gains. From short-term actions such as sending SMS messages to enemy soldiers to get them to desert, to long-term operations such as the "First Great Information War" that crushed the Soviets[8], the pace is able to vary in accordance with the conflict. Different conflicts require different strategies and approaches, and every conflict has its own logic.

Apo Sahagian, an Armenian volunteer from Jerusalem who fought in the Second Karabakh War, makes a critical remark about the pacing of these influence operations, despite not being educated in this field.[9] What he observed during the Karabakh War that made him resentful is that the path to defeat can happen if one side relies on long-term influence operations too much and neglects the crucial role of violence, which Gerasimov had also implied.

As Sahagian noted, the protestors in the major cities of the Western World could not save Hadrut, Armenian diaspora could not defend the Gates of Shushi from the Turks, and social media campaigns could not remove the Turks from Shushi.[10] These types of long-term influential means, which ultimately failed in this conflict, have been ongoing in Western media and politics for decades and did next to nothing to help the Armenian cause in the Karabakh conflict. Armenians sought a non-kinetic shield that would save them from the kinetic hammer of the Turks, but ultimately the hammer crushed the shield to dust.

Every conflict has its own logic, and conflicts such as the Syrian and Karabakh conflicts require more conventional kinetic means when compared to the famous *hybrid* conflict of the Ukrainian crisis of 2013. In both conflicts, the conventional methods of regular armies played the lead role, but the non-kinetic means had different levels of effectiveness contributing to the outcome. Both conflicts were more conventional compared to the Ukrainian war in 2013, but influence operations still played an essential role, which is the reason for writing this chapter.

The proposal is that quick-impact influence operations were much more effective than long-term operations in conflicts such as Türkiye's operations in Syria (e.g., Olive Branch and Peace Spring) and in the Karabakh War. A march in Los Angeles against Azerbaijan or social media campaigns perpetrated by the PKK against Türkiye did not help their war effort; however, Türkiye's quick-impact influence operations, conducted using drone footage, did cement the Turkish victories. Conflicts that require conventional muscle tend to proceed better with QIA.

The following section explains what is meant by QIA and why Türkiye should keep focusing on this approach. QIA has both its strengths and weaknesses, which will both be elucidated. Of course, the inevitable military change that has played a significant role in the Turkish approach is also discussed in the following section. The second section explores the real experiences of Turkish quick-impact influence operations. These operations have four important outcomes that will be elaborated in detail.

The Quick-Impact Approach

The end of the Cold War triggered a domino effect regarding various aspects ranging from political to military changes. The famous hypothesis of *revolution in military affairs* was seen to increase its pace after the Cold War, as Tolga Ökten and Emrah Özdemir have proposed in this book. This change was mainly kinetic in Türkiye, as other non-kinetic means were not focused on enough due to these being considered as *side ways* that are the products of *metis*, which has been considered a tool of the weak in the Western understanding of war.[11] Contrarily, this understanding has gradually changed due to the global approach to war becoming more diversified with the Russian[12] and Chinese approaches[13] gaining prominence, as well as due to Türkiye's personal experience in its conflict against the PKK.

The PKK extensively uses social media and conventional media to wage an influence war against Türkiye. In recent years, they have increased their efforts focusing on concepts trending in contemporary times. Concepts such as women's rights, democracy, pluralism, diversity, economic justice, and even environmental sustainability[14] appeal to the West, and PKK uses these to play the long game in the information sphere. PKK primarily targets feminists, as PKK has female fighters among their ranks. PKK deliberately promotes the so-called "modern" Kurdish female into a "freedom fighter," seeking long-term gradual support from the West by creating fake images that glorify their terrorist organization. Despite reaching a large audience on social media platforms, mainly YouTube, the actual effect these efforts have from a tangible perspective is debatable at best.

Creating an influence in the West sounds like a compelling game-winning concept and may appear like *supreme excellence*, but Jiyan[15], PKK's self-named female guerilla fighter, was unable to rally the Amazonian forces of Europe to her cause. Her life being presented as PKK's *Amazon fighter* may have influenced some people in Europe, but they probably turned off YouTube after the video, sipped their coffee, and went back to their everyday lives in Amsterdam or Berlin, remembering nothing about Jiyan or her cause after a few days. Such propaganda bears no fruit and can only serve as a preparatory phase for decisive influence operations.

Just like Jiyan's video, the *Her War: Women vs. ISIS* video that reached 2.5 million views on YouTube, had a description detached from reality: "The enemy fears female warriors. Jihadists believe if they are killed by a woman, they will go straight to hell".[16] No jihadists were running from these female warriors. It is the author's opinion that these types of social media engagements have no discernible effects on the conflict, they just mainly serve as a *reality show* for Western audiences.

As stated in the first paragraph of this section, PKK tries to present itself as a "friend to the people" and the "good guys." This approach was the same in Syria, but the truth is quite the opposite. Gökçen Yılmazlı, a feminist humanitarian worker in northern Syria, interviewed Turkmen women in the region. The Turkmen women stated how PKK maltreats all non-Kurdish women in the region. This alone falsifies PKK's feminist claims, because they only promote Kurdish women for their propaganda while other women suffer under their fists. Similar to the treatment of women, PKK's attacks on vital civilian facilities steeply declined after Turkish intervention in the region.[17] The Turkish Armed Forces (TAF) has reduced violence against both men and women in the region. Nevertheless, PKK conducts effective long-term influence operations, despite the truth being the opposite of their claims.

Such a long-term approach to influence operations has three main problems. First, the presentation of a fake image, such as in the PKK's example, can be challenged by providing the opposing proof over time. Secondly, even if the *truth* is used for an influence operation, it can be suppressed by the adversary's opposing disinformation operations. Never-ending debates in social media or other social circles caused by conflicting information bear no fruit. And lastly, when long-term operations improperly select their targets, a long time is required to understand the targeting failure and restructure the operations appropriately.

This is why QIA is more effective in conflicts in our times, at least in some conflicts such as the Syrian operations and the Karabakh War. When these QIA operations successfully reach their target, short-term objectives are achieved that cannot be undone. Neither falsification nor disinformation efforts can be effective in operations seeking such short-term goals. In addition to this, such quick operations work better in contemporary societies that have a relatively short attention span in the media and on the Internet.

Have PKK done anything correctly in terms of effective quick-impact influence operations? Yes, they have. During the Syrian crisis, PKK increased its efforts to influence leftist and anarchist groups in the West. They did achieve considerable short-term success. After forming the International Freedom Battalions (EOT) in June 2015, they attracted many different leftist groups from the West. The EOT was the umbrella organization defending Rojava against ISIS, and during its *hype* phase, the organization had a variety of groups such as the Greek anarcho-communist group known as the Revolutionary Union for Internationalist Solidarity, an English-speaking unit called the Bob Crow Brigade made up of British, Irish, and Canadian Leftists, and the International Revolutionary People's Guerrilla Forces (IRPGF).[18] At some point, PKK was even supported by The Queer Insurrection and Liberation Army (TQILA), as ISIS had been persecuting and murdering all homosexuals within their reach. TQILA had no military capabilities but did help with promoting PKK in the Western information sphere.[19]

Although the exact figures for these foreign fighters are unknown, PKK mustered a considerable number of them in the short term. Their quick-impact operations centered around the *Rojava* rhetoric were successful. This was due to their effective propaganda on social media and WhatsApp, which asked for help from the extreme leftist groups for the *revolution*.[20] Furthermore, these recruits would then serve PKK's long-term propaganda upon returning to their home countries by engaging in *peaceful activism*, which mobilizes Western audiences for further recruitment, thus killing two birds with one stone.[21]

One might argue that the long-term information operations criticized earlier have a cumulative effect with regard to the short-term propaganda, and this argument cannot be ignored. However, influence operations are complex and should be approached from a holistic perspective. Both QIA and the long-term approach have their uses. Therefore, the importance of long-term propaganda should not be neglected, but QIA operations are needed for concrete results. They need to be employed at the point when the conflict culminates: If states and non-states alike are unable to utilize these quick-impact operations at the focal point of the conflict, they will not be able to reap the benefits of long-term efforts.

Like PKK's influence war against Türkiye, Armenia and its diaspora have been engaged in a similar effort against Türkiye and Azerbaijan. The promotion of the so-called Armenian genocide could be one of the most extended information warfare operations waged upon Türkiye. Even though many countries recognize the events as a genocide, does this recognition serve anything concrete? The short answer is, no, it serves nothing relevant. All the recognizing countries still engage in trade, diplomacy, and all other functions that every normal country has with another.

During the Syrian intervention, the Armenian diaspora sparked the genocide talks once more. Trying to brand the Turkish intervention as a second genocide committed against Kurdish people, they sought in the information sphere to promote the so-called Armenian genocide and undermine the Turkish forces' legitimacy. What were the results? Nothing concrete, as noted already. Some liberals and left-leaning intellectuals condemned Türkiye on both social media and conventional media, but that was just talk that revealed nothing profound about the actual conflict. Thousands of retweets and millions of likes bring nothing to the field.

Though perhaps sounding strange, one example of what really counted in the Second Karabakh War was a song produced by the famous System of a Down group, known for its pro-Armenian stance as they are American Armenians. The music video for "Those Who Protect the Land" was released on November 6, 2020 to support the effort in the Karabakh War.[22] The slogan-like lyrics of "Would you stay and take a stand?" or the glorification of Armenian soldiers in the conflict with "They protect the land" profoundly affect the listener. Contrarily though, the song did not significantly raise the morale of Armenian forces, as they were unable to "protect the land" from the "scavengers and invaders" but instead deserted their armies. Neither Serj Tankian's inspirational voice nor the serious looks of Daron Malakian created a lasting effect on Armenian morale, but the song did have an effect in terms of quick-impact goals.

The song was successful both from a musical and video perspective, raising more than $275,000 for the war effort; this alone is quite an achievement gained from a single influence operation conducted by a music band. The influence aspect of the song did not help the conflict, but at least the money was a tangible form of aid. Such quick results serve much better than the century-old Armenian genocide story in terms of concrete effects. Aid raised from the song was for buying first-aid kits and other items for Armenian soldiers in need.[23] This impact mattered within the conflict, compared to the great pro-Armenia marches in the United States and other Western countries or the extensive social media campaigns that only served as a self-relieving instrument for the Armenian diaspora.

What worked in these two mentioned conflicts was QIA. Long-term plans in the information space only become able to bear fruit through these quick-impact operations.

QIA in influence operations is an approach that seeks short-term goals with concrete, tangible results. Through careful targeting and timing, such operations may yield more results in a conflict compared to never-ending information campaigns. Without these quick-impact operations, long-term influence operations mean almost nothing. Therefore, these quick-impact operations should aim for the conflict's center of mass to get quick results. These qualities and goals qualify QIA as a type of *war of annihilation* in the information sphere.

Armenia and its diaspora could have focused on consistent short-term operations rather than trying to exhaust Türkiye and Azerbaijan in the diplomatic and information spheres with their so-called Armenian genocide influence operation. Türkiye and Azerbaijan, however, are firm regarding the short-term game. Türkiye's non-kinetic QIA cemented the kinetic victory of Karabakh with no path of recourse. Armenia failed with its long-term approach, while Azerbaijan and Türkiye succeeded with their short-term approach.

What the above points out does not mean that long-term influence operations fail to deliver. With the example of the Cold War and the longest exhaustion influence war in history, an exhaustion strategy in an influence war is able to work, but the author believes that recent conflicts, being quicker in the information sphere and smaller in scale, require a QIA. Although both influence strategies have their uses in different conflicts, there is no silver bullet in the influence war.

As seen in Figure 9.1, QIA actions are positioned in the highly effective, short-term area. The positionings of the operations are not based on exact numbers but rather provide the reader with a practical understanding. It is proposed that *violence-centric operations* with limited aims, such as the Second Karabakh War and the Turkish operations in northern Syria, require quick influence operations (IO). They do not require a complex influence campaign to affect the adversaries in the long term, but the operations should instead act as a tool to support the conventional military operations in the short term.

Different operations with quick positive outcomes might be positioned in the QIA zone marked in the figure, such as footage of Turkish drones, PKK's

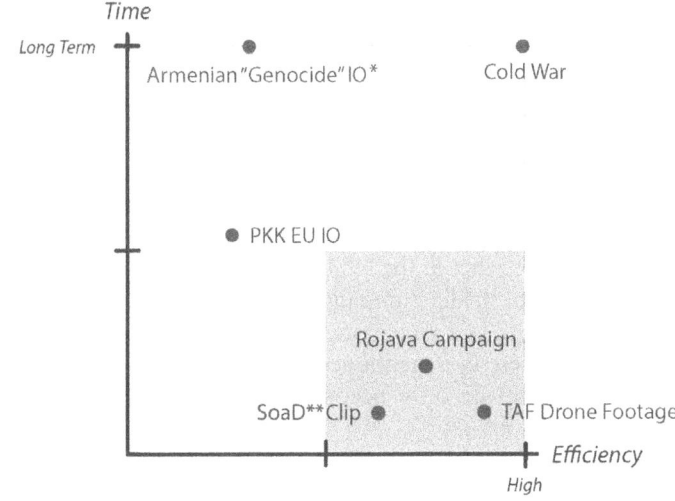

Figure 9.1 The Quick-Impact Approach.

Sources: *Created by the author.*

Rojava propaganda targeting Europe, and even the System of a Down music clip that provided financial support for the Armenian soldiers. The QIA zone represents getting meaningful positive results through short-term IOs.

On the flip side, as seen in the Cold War example, long-term IOs may provide victory, but achieving one's goal is much harder due to the reasons mentioned in this section. Winning the Cold War required two giants to battle each other in every possible area, including the information sphere. Long-term operations require a clear goal, extensive planning, an inclusive information strategy, and resources. PKK's IOs that targeted Europe and the so-called Armenian Genocide IOs fell short of delivering victory, as such efforts were beyond their power and capacity and lacked the international audience's interest. QIA does not require such an extensive endeavor: It is easier to conduct and collect benefits in the short run.

The next section, clarifies the definition of QIA, as both TAF's experiences in Syria and Azerbaijani Armed Forces' (AAF) experiences in Karabakh provide clear examples of QIA.

The Turkish Experience

Influence operations, especially long-term ones, require a wide range of experts, a holistic approach, and most importantly, an efficient centralized leadership, usually led by the civilian government itself. The Turkish Armed

Forces (TAF) does not possess the extensive capabilities to run influence operations independently. This lack of capacity is not just about TAF but also about the army structure itself: Neither in Russia nor in the United States can the armies take up the challenge of such complex information warfare.

Russians have been considered masters of manipulation and dis/misinformation in recent years. The Kremlin leads the effort in influence operations alongside the FSB and GRU, but the so-called soldiers of the influence war have been the civilians used as the trolls, hackers, and pawns campaigning through social media. Even if the army were to lead the influence war, the employed tactics would be executed by civilians. The military personnel can neither conduct such micro-operations nor do they want to be held responsible for these operations. How can an army invent memes or troll adversaries every day to create an indirect long-term effect on their adversary's people? Simply put, they cannot.

Were the United States Armed Forces able to undermine the Soviet ideology and integrity of its people on their own? No, a holistic effort on the part of the United States, its institutions, experts, people, and army, as well as all other indirect cultural and sociological aspects was what took down the "Red Menace" in the Cold War.

What armies and TAF in the Turkish context can do on their own, though, are the quick-impact operations that seek quick, direct, short-term results. Thus, TAF should focus on what it is capable of without getting into the complexity and extensiveness of the influence war. Drones, especially the Turkish use of drones[24], have been quite the military innovation in recent years, affecting tactical, operational, and strategic levels. Some scholars view drones as mere equipment that does not change how war is conducted; however, the debate has become livelier than ever, as drones played a key role in the Karabakh War.

For example, Can Kasapoğlu argued that the success of Turkish drones was not just about the equipment per se, but about the way they are used and how Turkish forces put them into play.[25] Drones may or may not trigger a doctrinal change in the military, but this debate is not the concern here, due to drones already being the subject of other chapters in this book. Instead, this chapter is concerned with a focus on the *side-effects* of drones from a non-kinetic perspective.

If Russians were perfecting their so-called hybrid warfare capabilities from the early Chechen war until the late Ukrainian crisis, then Turks have been perfecting their use of drones in recent conflicts, regarding both kinetic and non-kinetic uses. From an operational standpoint, Turkish forces skillfully used their drones throughout the beginning of the Syrian war and the short Karabakh war. These drones were also used in the quick-impact influence operations through social media and conventional media spheres. In fact, the promotion of these drones in the information sphere was so effective that the Armenian side showed a very similar response by releasing footage of Turkish drones being shot down.[26]

The high-quality and high-resolution drone footage was used by TAF for one primary and three secondary purposes. First, TAF used this to demoralize enemy soldiers; this is QIA's most critical aspect, because desertion and low army morale play a crucial role in military power. The first secondary purpose was to demoralize public opinion of the adversaries through the drone footage presented on social media. The second secondary purpose was to use the footage to consolidate the internal information sphere and increase support for the military operations. Last but not least, the footage was used for fact-checking and reducing the fog of war.

The actual results from these quick influence operations were that this footage was able to become one of the main driving factors causing enemy soldiers to desert. If the information on the Internet affects the cohesion of Russian units,[27] it can also cause Armenian soldiers to break ranks and run away. Nearly every soldier uses a smartphone during a conflict, even when they are banned during an operation. Reaching the hearts and minds of soldiers is easier in the digital age.

During the Karabakh war, Armenian soldiers deserted their posts in massive numbers. Even though the actual number is quite shaky at best, 10,000 Armenian soldiers were estimated to have deserted the army and to have been court-martialed.[28] Desertion has many complex reasons that cannot be fully explored in this chapter which aims only to underline the effects from drone footage.

Drones were one of the most critical pieces of equipment affecting the outcome of the Karabakh War, both with their kinetic and non-kinetic uses. Some experts such as Kasapoğlu went even further and claimed what had sealed the deal in the war was not just the drones that Turkish forces used but also how they were used.[29] The effective use of these drones and the kinetic damage they caused were remarkable and proved to be a demoralizing factor for the enemy units. Furthermore, it seems that the non-kinetic damage these drones caused was also quite relevant to breaking the will of the Armenian soldiers and PKK-KCK terrorists.

The footage of Armenian soldiers and equipment being easily hunted down by the Turkish drones increased the desertion of Armenian soldiers. From images of crying deserting Armenian soldiers to destroyed Armenian vehicles and cheerful Turkish forces, both Azerbaijan and Türkiye used videos effectively on social media and reached high view counts. The use of drone footage was the last nail in the coffin of the already failing Armenian morale and proved to be an effective means in the information sphere.

The first secondary effect from this footage involved public opinion and support for the war abandoning the enemy. The Armenian people were already divided among themselves prior to the war. Some accused Nikol Pashinyan of being a puppet of the West. By leaning too much on the West (i.e., France) and leaving the old alliances aside, the Armenian people were afraid of Russia not helping them in the wake of an Azerbaijani advance. More or less, their fear has become a reality, as Russia made no move until the last moment of the conflict, and that was just to protect Russian interests in the region and teach the Armenian leadership a lesson.

As the Armenian people were divided into at least two (i.e., those for and against Pashinyan) and the war started with catastrophes on the Armenian side, the Armenian people quickly descended into despair. The people's already failing morale was also being bombarded by the drone footage and other videos that were presenting the Turkish army as the victor. The image was clear: The war was already over. The footage served as simple "proof" of this reality. With the news coming from the front enhanced by this footage, Armenian people also joined the ranks of desertion, and their support for the war dropped.

The second point is the opposing symmetry used to shatter the enemy's will: the consolidation of the internal public sphere and increased public support for the ongoing operations. During the Syrian Operations and the Karabakh War, the footage raised public support for the Turkish forces. The drone footage and pro-Turkish videos reached millions of views combined from YouTube and Twitter. For example, the official YouTube channel of the Ministry of Defence of the Republic of Azerbaijan is *Azerbaycan Respublikası Müdafia Nazirlıyı* and presented lots of videos about the Armenian forces being hit by drones in the Karabakh War. Eight of the top ten most-watched videos on the channel contain drone footage about destroyed Armenian forces. Most of these views (all the drone footage obtained more than 500,000 views) can safely be assumed to belong to the Turkish audience. Therefore, one might argue that these videos served as an effective means for reinforcing the people's support for the Karabakh War.

These "war has already been won" videos increased the legitimacy of Turkish forces in Northern Syria as well. According to polls conducted in 2016, the Turkish support for a potential Syrian operation was 61%.[30] However, with the increased effort to promote the Turkish populace's support for the Syrian operations, the support rate increased to between 85% and 90% in different polls during the operations.[31] Thus, the Turkish government ideally consolidated the public space, and the Turkish public believed that these operations had a just cause. Of course, this amount of high support cannot be explained only by the quick-impact means, as the Turkish government used all types of media in this endeavor, but QIA operations did serve the Turkish cause to raise public support.

The last point, which is the fact-checking property of Turkish quick-impact operations is crucial, as this can also be used in lawfare. From the start of the Syrian conflict, PKK accused Turkish forces of targeting civilians, forcing them out of their homeland, and committing crimes against humanity. TAF focused on repudiating these claims with their drone footage.

One example opposing PKK's claims was caught by Turkish drones. According to the drone footage, PKK terrorists are spotted blocking Kurdish civilians trying to get out of Afrin.[32] They did not let any Kurdish civilians out of the city. A long convoy was seen trying to leave the city, but none achieved their goal because PKK terrorists forcibly kept them inside the city for their political goals. They aimed to keep the Kurdish civilians within the city while forcing other ethnicities out of the city to claim Afrin as a *pure* Kurdish city, thus giving PKK legitimacy over the region. Acting contrary to what they accused Türkiye of, they were themselves caught red-handed by the drones.

Another example is the footage TAF took that clearly shows civilians being targeted by PKK artillery. Even though the terrorists burnt tires to block drone footage, three terrorists were plainly visible, shooting rockets at the Turkish towns. In 24 days, these rocket attacks killed two Syrian and five Turkish civilians, and caused 113 to be injured.[33]

In Syria, PKK tried to create an image of Türkiye as an invader who violates fundamental human rights and targets civilians. The terrorists prepared fake footage and propaganda to achieve this goal.[34] Contrary to the PKK terrorists' claim, they were the ones who targeted civilians during the Turkish operations in Syria, as pointed out in earlier paragraphs. Although Türkiye could not capitalize on the truth in the international media, TAF did its best to spot PKK's transgressions against Turkish and Syrian civilians. The drone footage presented to the Turkish news agencies could not reach a broad international audience, but the truth is there for the ears who want to hear it.

The one primary and three secondary purposes of the Turkish QIA operations mentioned above show that quick influence campaigns should support kinetic successes. Rather than focusing on long-term operations, some conflicts require QIA. The Turkish experiences in northern Syria and Karabakh show that armies should focus on these approaches yielding fast results as armies lack the capacity to conduct extensive influence campaigns. Armies should focus on what they do best.

Conclusion

Non-military means affecting military change is a paradoxical concept. Yet, as stated before, influence operations have become more and more effective in the so-called hybrid war. These influence operations became a complex and extensive effort, and thus no army could undertake this challenge alone. It requires a full-state approach that concerns itself with both conventional and new influence spaces. What armies can do on their own are quick-impact operations.

Türkiye lacks a coherent and holistic influence strategy as a state. For years, neglect and political agendas have undermined Turkish influence strategies. Being stripped of institutions such as the Psychological Warfare Division (PHD), Türkiye is vulnerable in the information sphere. On the other hand, the Turkish Armed Forces continues to be the only effective institution in the information sphere. TAF's quick-impact approach (e.g., using drone footage, pro-Turkish videos) served as an effective tool against Türkiye's opponents in various aspects, as mentioned in the previous sections, as well as consolidating the populace.

Türkiye's strong suit might not be long-term influence campaigns, but TAF provides a different solution to the problem. Some conflicts require quick solutions rather than never-ending influence campaigns. As stated, the century-old so-called Armenian Genocide influence operation yielded no results in the Karabakh War, but simple drone footage did help the Turkish war effort. Every conflict has its own logic, and every influence conflict has its own logic as well.

As a closing remark, I should state that the effectiveness of QIA requires extensive research. I am aware that the chapter presented here is not scientific in terms of quantitative research, but it does serve as a thought exercise. I advise further researchers to focus on computational methods to research the effectiveness of TAF's social media actions as well as sociological methods such as interviewing the deserters of the Syrian conflict and the Karabakh War. Unfortunately, I fall short regarding both my recommendations as I am not an expert on computational methods, nor do I possess the qualities to interview Armenian soldiers or PKK terrorists.

Notes

1 Azar Gat, *The Causes of War and the Spread of Peace: But Will War Rebound?*, Oxford University Press, 2017.
2 "Top Russian General Lays Bare Putin's Plan for Ukraine", *HuffPost*, https://www.huffpost.com/entry/valery-gerasimov-putin-ukraine_b_5748480. accessed on November 22, 2020
3 Carl von Clausewitz, Michael Howard, and Peter Paret, *On War*, Princeton University, 1976. p. 75
4 "Top Russian General Lays Bare Putin's Plan for Ukraine", HuffPost, https://www.huffpost.com/entry/valery-gerasimov-putin-ukraine_b_5748480, accessed on November 22, 2020
5 Gat, *The Causes of War and the Spread of Peace: But Will War Rebound?*, p. 244.
6 Edward N. Luttwak, "Toward Post-Heroic Warfare", *Foreign Affairs*, 74:3, 1995, p. 109–122.
7 Sun Tzu, and J. Minford, *The Art of War*, Penguin Books, New York, 2008, p. 12.
8 Jolanta Darczewska, "The Anatomy of Russian Information Warfare", *Point of View*, 42, 2014, p. 15.
9 "Our Useless Diaspora, Our Future Armenia", *The Armenian Weekly*, https://armenianweekly.com/2021/02/03/our-useless-diaspora-our-future-armenia/, accessed on November 22, 2020.
10 "Our Useless Diaspora, Our Future Armenia", *The Armenian Weekly*. https://armenianweekly.com/2021/02/03/our-useless-diaspora-our-future-armenia/, accessed on November 22, 2020.
11 Jean-Vincent Holeindre, *La Ruse Et La Force: Une Autre Histoire De La Stratégie*, Perrin, Paris, 2017, p. 15–16
12 Ofer Fridman, *Russian Hybrid Warfare: Resurgence and Politicisation*, Oxford UP, New York, 2018.
13 Qiao Liang, and Wang Xiangsui. *Unrestricted Warfare*, PLA Literature and Arts Publishing, Beijing, 1999.
14 Rana Khalaf, "Governing Rojava Layers of Legitimacy in Syria", *Chatham House Research Paper*, Dec 2016.
15 "Jiyan: Story of a Female Guerilla Fighter", *Youtube*, https://www.youtube.com/watch?v=oEQEidgcMhU, accessed November 1, 2021.

16 "Her War: Women vs. ISIS", *Youtube*, https://www.youtube.com/watch?v= uqI0a4VgEs8, accessed November 1, 2021.

17 "The Most Notable Human Rights Violations in Syria in October 2020", *SNHR*, 2020, https://snhr.org/wp-content/pdf/english/The_Most_Notable_ Human_Rights_Violations_in_Syria_in_October_2020_en.pdf, p. 16 accessed November 1, 2021.

18 "Syrian Kurds Announce Unified Foreign Fighter Unit", *Kyle Orton's Blog*, 2015, https://kyleorton.co.uk/2015/06/13/syrian-kurds-announce-unified-foreign-fighter-unit/, accessed on November 1, 2021.

19 "The Latest Chapter of Syria's Media War: A 'Gay Unit' Fighting the Islamic State", *Kyle Orton's Blog*, 2017 https://kyleorton.co.uk/2017/07/31/the-latest-chapter-of-syrias-media-war-a-gay-unit-fighting-the-islamic-state/, accessed on November 1, 2021.

20 "European Far-Left Fighters in Syria: Take Them Seriously", *Euronews*, 2018, https://www.euronews.com/2018/03/05/european-far-left-fighters-in-syria-it-s-time-to-take-them-seriously-view, accessed November 1, 2021.

21 "Far Left on the Front Lines: The Westerners Joining the Kurds' Fight in Syria" *France 24*, 2018. https://www.france24.com/en/20180223-syria-afrin-foreigners-westerners-far-left-join-kurdish-revolution-fight-turkey, accessed November 1, 2021.

22 "System of a Down – Protect the Land (Official Video) – *Youtube*", https://www.youtube.com/watch?v=XqmknZNg1yw, accessed on November 7, 2021.

23 "System of a Down Live Fundraiser for Wounded ... – *Youtube*", https://www.youtube.com/watch?v=1q05uG8Lag8, accessed on November 7, 2021.

24 By Turkish, I mean both Azerbaijan and Turkey. This is an ideological and cultural choice of mine as both nations consider themselves "Turkish". The word "Turkic" was not used on purpose as it is a fabricated term to denote Turkishness of other Turkish republics such as Kazakhstan, Uzbekistan etc.

25 "Analysis – Heavily-Armed and Dangerous: The Akinci Drone in the Skies", *Anadolu Ajansı*, https://www.aa.com.tr/en/analysis/analysis-heavily-armed-and-dangerous-the-akinci-drone-in-the-skies/2357792, accessed on November 9, 2021.

26 "Հերթական Աքս-ների Խոցումները. Արցախի ՊԲ-ի Հաստակյութը – *Youtube*, https://www.youtube.com/watch?v=6l_ryQptb9k, accessed on November 8, 2021.

27 "Russia bans soldiers from using smartphones and tablets over OPSEC concerns", *Radio Free Europe*, https://taskandpurpose.com/news/russia-bans-smartphones-tablets-soldiers/, accessed on November 15, 2021.

28 "Armenia Launches Criminal Cases against 10,000 Soldiers for Desertion in War." *AzerNews*, https://www.azernews.az/karabakh/176093.html, accessed on November 3, 2021.

29 "Analysis – Heavily-Armed and Dangerous: The Akinci Drone in the Skies", *Anadolu Ajansı*, https://www.aa.com.tr/en/analysis/analysis-heavily-armed-and-dangerous-the-akinci-drone-in-the-skies/2357792, accessed on November 9, 2021.

30 "Adil Gür Suriye Anketi Sonucu Operasyon Olursa...", *Internet Haber*, https://www.internethaber.com/adil-gur-suriye-anketi-sonucu-operasyon-olursa-1566477h.htm, accessed on November 12, 2021.

31 "Mak Danışmanlık: Halkın Yüzde 85'I Afrin Operasyonunu Destekliyor." *Sputnik Türkiye*, https://tr.sputniknews.com/20180129/mak-danismanlik-anket-afrin-operasyonu-1032011906.html,accessed on November 12, 2021.

32 "Afrin'den Çıkmak Isteyen Sivilleri Tehditle Durdurdular", *Anadolu Ajansı*, https://www.aa.com.tr/tr/dunya/afrinden-cikmak-isteyen-sivilleri-tehditle-durdurdular/1086593,accessed on November 12, 2021.

33 "Türkiye'ye Top Atışı Yapan Teröristler Vuruldu! En Net Görüntüler", *Hürriyet*, https://www.hurriyet.com.tr/gundem/turkiyeye-roket-atan-teroristler-ilk-kez-bu-kadar-net-goruntulendi-40740503, accessed on November 20, 2021.

34 "Terör Yandaşlarından Sosyal Medyada Üç Yalan Daha", *Anadolu Ajansı*, https://www.aa.com.tr/tr/gunun-basliklari/teror-yandaslarindan-sosyal-medyada-uc-yalan-daha-/1039728, accessed on November 1, 2021.

References

"Adil Gür Suriye Anketi Sonucu Operasyon Olursa...", *Internet Haber*, https://www.internethaber.com/adil-gur-suriye-anketi-sonucu-operasyon-olursa-1566477h.htm, accessed on November 12, 2021.

"Afrin'Den Çıkmak Isteyen Sivilleri Tehditle Durdurdular", *Anadolu Ajansı*, https://www.aa.com.tr/tr/dunya/afrinden-cikmak-isteyen-sivilleri-tehditle-durdurdular/1086593, accessed on November 12, 2021.

"Analysis – Heavily-Armed and Dangerous: The Akinci Drone in the Skies", *Anadolu Ajansı*, https://www.aa.com.tr/en/analysis/analysis-heavily-armed-and-dangerous-the-akinci-drone-in-the-skies/2357792, accessed on November 9, 2021.

"Armenia Launches Criminal Cases against 10,000 Soldiers for Desertion in War", *AzerNews*, https://www.azernews.az/karabakh/176093.html, accessed on November 3, 2021

Clausewitz, Carl von, Howard, Michael and Paret, Peter, *On War*, Princeton University, Princeton New Jersey, 1976.

Darczewska, Jolanta, "The Anatomy of Russian Information Warfare", *Point of View 42*, 2014.

"European Far-Left Fighters in Syria: Take Them Seriously", *Euronews*, 2018. https://www.euronews.com/2018/03/05/european-far-left-fighters-in-syria-it-s-time-to-take-them-seriously-view, accessed on November 1, 2021.

"Far Left on the Front Lines: The Westerners Joining the Kurds' Fight in Syria", *France 24*, 2018. https://www.france24.com/en/20180223-syria-afrin-foreigners-westerners-far-left-join-kurdish-revolution-fight-turkey, accessed on November 1, 2021.

Fridman, Ofer, *Russian Hybrid Warfare: Resurgence and Politicisation*, Oxford University Press, New York, 2018.

Gat, Azar, *The Causes of War and the Spread of Peace: But Will War Rebound?* Oxford University Press, Oxford, 2017.

"Her War: Women vs. ISIS", *Youtube*, https://www.youtube.com/watch?v=uqI0a4VgEs8, accessed November 1, 2021.

Holeindre, Jean-Vincent, *La Ruse Et La Force: Une Autre Histoire De La Stratégie*, Perrin, Paris, 2017.

"Jiyan: Story of a Female Guerilla Fighter", *Youtube*. https://www.youtube.com/watch?v=oEQEidgcMhU, accessed on November 1, 2021.

Khalaf, Rana, "Governing Rojava Layers of Legitimacy in Syria", *Chatham House Research Paper*, December 2016.

"Mak Danışmanlık: Halkın Yüzde 85'I Afrin Operasyonunu Destekliyor", *Sputnik Türkiye*, https://tr.sputniknews.com/20180129/mak-danismanlik-anket-afrin-operasyonu-1032011906.html, accessed on November 12, 2021.

"Our Useless Diaspora, Our Future Armenia", *The Armenian Weekly*. https://armenianweekly.com/2021/02/03/our-useless-diaspora-our-future-armenia/, accessed on November 22, 2020.

Liang, Qiao and Xiangsui, Wang, *Unrestricted Warfare*, PLA Literature and Arts Publishing, Beijing, 1999.

Luttwak, Edward N., "Toward Post-Heroic Warfare", *Foreign Affairs*, 74:3, 1995, p. 109–122.

"Russia bans soldiers from using smartphones and tablets over OPSEC concerns", *Radio Free Europe*, https://taskandpurpose.com/news/russia-bans-smartphones-tablets-soldiers/, accessed on November 15, 2021.

"Syrian Kurds Announce Unified Foreign Fighter Unit", *Kyle Orton's Blog*, 2015, https://kyleorton.co.uk/2015/06/13/syrian-kurds-announce-unified-foreign-fighter-unit/, accessed on November 1, 2021.

"System of a Down - Protect the Land (Official Video)", *Youtube*. https://www.youtube.com/watch?v=XqmknZNg1yw, accessed on November 7, 2021.

"System of a Down Live Fundraiser for Wounded", *Youtube*. https://www.youtube.com/watch?v=1q05uG8Lag8,accessed on November 7, 2021.

"Terör Yandaşlarından Sosyal Medyada Üç Yalan Daha", *Anadolu Ajansı*. https://www.aa.com.tr/tr/gunun-basliklari/teror-yandaslarindan-sosyal-medyada-uc-yalan-daha-/1039728, accessed on November 1, 2021.

"The Latest Chapter of Syria's Media War: A 'Gay Unit' Fighting the Islamic State", *Kyle Orton's Blog*, 2017 https://kyleorton.co.uk/2017/07/31/the-latest-chapter-of-syrias-media-war-a-gay-unit-fighting-the-islamic-state/, accessed on November 1, 2021.

"The Most Notable Human Rights Violations in Syria in October 2020", *SNHR*, 2020, https://snhr.org/wp-content/pdf/english/The_Most_Notable_Human_Rights_Violations_in_Syria_in_October_2020_en.pdf, p. 16, accessed on November 1, 2021.

"Top Russian General Lays Bare Putin's Plan for Ukraine", *HuffPost*, https://www.huffpost.com/entry/valery-gerasimov-putin-ukraine_b_5748480, accessed on November 22, 2020.

"Türkiye'ye Top Atışı Yapan Teröristler Vuruldu! En Net Görüntüler", *Hürriyet*, https://www.hurriyet.com.tr/gundem/turkiyeye-roket-atan-teroristler-ilk-kez-bu-kadar-net-goruntulendi-40740503, accessed on November 20, 2021.

Tzu, Sun, and Minford, J., *The Art of War*, Penguin Books, New York, 2008.

"Հերթական Ապս-Ների Խոցումները. Արցախի ՊԲ-ի Տեսանյութը", *Youtube*. https://www.youtube.com/watch?v=6l_ryQptb9k, accessed on November 8, 2021.

Conclusion

The Manifest and Latent Sources of Turkish Military Innovation

Barış Ateş

Introduction

This research sheds light on a number of key aspects of Turkish military innovation and provides the opportunity to compare general military innovation research by providing a case study from outside the developed Western world. First of all, the fact that the interest in Türkiye's military innovation came to the fore only after its successes on the battlefield makes comprehending its background and past trials a challenge. Therefore, this study is a considerable first step in understanding the background of Turkish military innovation. Second, the study demonstrates how military innovation may be accomplished in a developing country that faces a variety of threats. Last, it explores the role coercive arms embargoes and restrictions have had on the innovation process.

Türkiye's long-term partnership with the West, in common trade, and in its institutions having been designed according to the Western model all lead to comparable outcomes. In other words, a development that occurs in the West may also eventually be experienced in Türkiye. To argue that this common denominator is greater with regard to armies would be no exaggeration. With the Turkish military having been a member of NATO for 70 years and one that has participated in several peacekeeping operations and adapted its essential components, such as organizational structure, doctrine, and weapons systems to NATO, developments in the West should, of course, affect the Turkish military. As a result, the objective of building an agile, professional, and high-tech army has been just as vital for Türkiye as it has been for other NATO members. Put simply, the wave of change that emerged in the West with the end of the Cold War led to a similar situation in Türkiye as well. However, the scene in Türkiye was a bit different. Of course, Türkiye was also relieved of a conventional threat and adapted some of the organizational changes that NATO pushed forward. However, terrorist groups had become a threat to the country's territorial integrity. Moreover, this threat coincided with the period when the West was experiencing the most comfortable years of the post-Cold War era. Consequently, even if Türkiye had wanted to follow the West in its military innovation movements, it was unable to.

DOI: 10.4324/9781003327127-11

Another difference concerned the capacity of a country that lacked sufficient economic power and scientific knowledge to produce systems able to trigger military innovation. This situation turned out to be a dependency relationship, and procuring weapons from NATO allies was the priority. However, this dependency took away Türkiye's freedom to act due to embargoes, restrictions, and associated political issues. In other words, achieving the means and capabilities to eliminate existing threats was no simple task. This situation led to frustration, and the reactionary response resulted in the revival of a state tradition that had developed the means and ability to do business under challenging conditions. Namely, the conditions pushed the country to innovate. From this perspective, that the *constraints have been the manifest source of innovation* in Türkiye can be safely confirmed. However, understanding which factors triggered this innovation is important, and the analyses in this book attempt to reveal the main characteristics.

The Factors Driving Turkish Military Innovation

The end of the Cold War provided unprecedented opportunities for military innovation enthusiasts, because the obvious changes were easily observable and provided the data needed for analysis. A robust literature can be said to exist in this field, especially regarding developed Western countries. Consequently, the driving factors of innovation movements, the relationships between them, their scope, and scale have been the subject of extensive research. Therefore, although no agreed-upon list is found in the literature, technology, operational environment, civil-military relations, alliance politics, domestic politics, cultural norms, and leadership come to the fore.[1] However, as for the period this book examines, the reasons behind the Turkish military innovation and how this innovation has been carried out reveal a rather intricate landscape. At first, the perceived risk and operational environment seem to be the main variables. However, upon considering that the same threat had been on the agenda in the past, the need to also take other factors into account becomes clear.

The interest in Turkish military innovation emerged not with the development of new systems but with their use, thus making the issue ambiguous and challenging to understand. The production of UAVs, missile systems, helicopters, and warships using national technologies attracted the attention of certain international market players from the very beginning. But, with the effective use of UAVs in conflicts such as Syria, Libya, and Karabakh in particular, Türkiye went beyond attracting attention to becoming the focus of international attention. In this case, military innovation was perceived as a technology-driven process. The present study reveals that this perception, while correct up to a point, is incomplete. UAVs have become a force multiplier and strengthened Türkiye's hand in many areas, from counter-terrorism to cross-border operations. However, the current events cannot be understood without looking at their past, in particular their philosophical background.

Technological developments can be seen as the primary determinant of military innovation. According to the approach known as technological determinism, technology and the scientist-entrepreneurs who produce it play the main roles. The ideas for a new weapon system have even been stated to initially have come not from soldiers but from groups of scientists and technologists. However, those who oppose this view argue that the success or failure of certain designs is determined by the social networks created by the military, political, and business elites.[2] As Latour alternatively emphasized, expanding networks and alliances are required for technology to succeed.[3] While explaining Turkish military innovation, Körpe referred to this concept of technological determinism in Chapter 1. However, he also emphasized that technology should be solution-oriented, or put more concisely, applicable. In his view, the geography of the constant crisis in which Türkiye exists and the accompanying material deficiencies or *absence* have caused Turks to incorporate technology into their lives with a pragmatic approach. As such, a desire exists to embrace technology. Dealing with defense technologies as one of the areas where this situation most commonly occurs, Körpe naturally determined technological factors to be more dominant than other factors. In this case, innovations in other fields caused by technology did not encounter any serious obstacle. The author explains this determination through the way the supply-demand relationship between defense industry companies and military forces has gained a symbiotic nature in recent years. Original designs that are able to affect strategic operations will be the tools essential for enabling this symbiotic relationship to turn into a breakthrough.

The production processes of defense technologies also support this symbiosis. As in other countries, technology experts in Türkiye produce some systems and place them under the service of the army. However, what makes Türkiye a bit different in this regard is that the technology manufacturer is present on the field alongside the users (i.e., the soldiers) throughout the entire production process. As Çam discussed in Chapter 3, the Bayraktar family is observed to have assumed such a role for UAVs. However, the family should also be noted to have endured every imaginable difficulty on this point by spending time at military bases in at-risk areas in order to develop their UAV. Therefore, the question of who has originated the above-mentioned technological innovation loses its meaning. This is because, while the scientist-entrepreneur group may have taken the initiative, a harmonious social network was also established at the same time due to this initiative being carried out with the user-level soldiers in the form of a bottom-up approach from the very beginning, as well as receiving support from top-level leaders.

As the threat increased, the soldiers already desired and expected high technology weapons systems. They did not hesitate to cooperate with civilian experts who promised success in this area. The development of the UAV took place in such a setting. The most important result from this cooperation was the absence of any resistance from the soldiers in terms of its use on the battlefield. In this way, the battle-tested UAV eventually became an enabler for changing the military doctrine and organization.

But was technological innovation only initiated by civilian experts? Eker answers this question in Chapter 4, while telling the story of the national warship project (MİLGEM). This project resembles the UAV process, in which Türkiye managed to produce a warship for the first time with its own means. More importantly, the software of the ship was developed by the soldiers and civilian engineers. Almost all the military engineers who were the architects of the project were Naval Academy graduates who had completed their postgraduate education at prestigious universities in the country and abroad. Yet the officers did not hesitate to cooperate with civilians in this work, which eliminated the stove-pipe approach and ultimately resulted in the project's success. To summarize, the soldiers who feel the threat closely and fight for their survival in their homeland definitely desire new technology for their armies. However, technology's role can be defined as that of *facilitator*. To explain military innovation through technological determinism becomes a shallow argument in this case. As sociologists have stated, the existence of strong social networks is seen as the primary factor that brings about success.

Another example showing technology alone to be insufficient concerns the size of the defense industry and the investments made in this sector. Öz shows the development of R&D investments in Chapter 2, where he examines the defense industry and the defense budget over the last 30 years. Thus, while interest in technology was present, developing it was impossible without the economic strength needed to realize it. Moreover, this period cannot be defined only through the establishment of military production factories and increased R&D expenditures. In addition, thousands of specialists were trained thanks to the investment in human resources. This group was educated at universities abroad or within Türkiye and became the locomotive for defense industry projects. Another situation that arises regarding this point concerns how the discontent and perhaps even anger created by the embargoes and restrictions had triggered a strong motivation toward innovation. What is striking here is that the political and military leaders who are in the position of decision-makers today came from a generation that had closely experienced the terrorist threat and knew the problems. Therefore, to state that enthusiasm for innovation exists rather than resistance against it would be no exaggeration. Thus, the military embargoes that started in the 1970s ignited a mindset at the top level that knew Türkiye also had to be self-reliant.

The military has had to innovate its organizational structure, doctrine, and even culture. Ökten demonstrates how the new technology helped change the maneuvers and use of firepower in Chapter 7. Of course, these changes were made possible to some extent by the UAVs. But in order to turn the opportunities UAVs provided into an advantage, a change was needed in the doctrine and effective professional units in the field. The number of these troops has nearly tripled over the last decade, and they are now equipped with small, flexible, advanced equipment and weapons systems. In this case, both the scope and speed of operations have increased and reached a level where terrorist organizations are no longer able to keep up. In short, technological improvements can be effective only when doctrinal and organizational changes accompany them.

Making a similar observation, Özdemir describes counter-insurgency operations as a solution frequently put on the agenda in times of crisis by comparing American, British, and Turkish examples in Chapter 6. According to Özdemir, innovation is not limitable to technological progress. On the contrary, it is a factor that can and should encourage changes in strategy, and organizational structure. As such, Özdemir states that the need exists for an organizational culture that is open to creative thinking and change. In this case, he views military innovation as a continuous process of re-evaluation and renewal. In short, technology is not a determinant on its own but is accepted as a facilitator.

Another explanation that strengthens this approach came from Öğün, who highlights the change in the international system in Chapter 5. Secondary states had depended on the alliance of which they were members in the bipolar world order and gained more freedom to act in the multipolar world order. When considering Türkiye as a typical example of this situation, the author examines how the emulation process transformed into an innovation process through the encouragement of the international system. Türkiye turns to reinvention not within the scope of invention but within the scope of innovation in proportion to its capacity. This situation is observed in Türkiye's use of weapons, which it has adapted to its body by imitating from other states, along with a unique military method. Regarding this point, Öğün focuses on the use of UAVs against a conventional enemy with swarm tactics and how it resultingly changes the balance on the battlefield.

The fact that this example actually changed the balance in Karabakh, Libya, and Syria increases the likelihood of imitation. The point to draw attention to here is the way the technology is used rather than its acquisition. As a matter of fact, the author defines this as a reinvention and states the main difference to be the way UAVs are used rather than their production. The author has proven that sociologists who emphasize the importance of the effective use of new technology rather than its production to be a reasonable claim. This new method exceeded the standard role UAVs have against limited and individual targets by revealing the use of drone swarms against armored units. Therefore, the harmony of social networks formed by the combination of different civilian and military groups has been reflected on the battlefield. The author foresees that, in accordance with the innovation cycle, other states will imitate and adopt Türkiye's military innovation.

The easily observable and measurable dimensions of military innovation are often the more prominent. Hence, kinetic abilities are emphasized. However, today's wars take place in front of the whole world thanks to the opportunities of the information age, and this requires armies to develop their non-kinetic capabilities. Dağhan Yet reveals in Chapter 9 how the military uses these abilities. However, the interesting point here is the dependence non-kinetic abilities have on kinetic ones. The *quick impact approach* (QIA) offers a way to dominate information warfare and is only applicable when drones are on the battlefield. The Turkish Army is seen to have sufficient experience in this regard. The traces of this experience can be seen in their counter-terrorist operations and in the conflicts in Syria, Iraq, Libya, and Karabakh. However, this is evidently not enough. Ultimately, information warfare should be able to be used independently from any kinetic effect. However, no information warfare capability can be easily said to have fully formed enough to counter the disinformation campaigns of terrorist organizations or other countries. In a nutshell, although the military and some civilian institutions are struggling *to the best of their ability* in a non-kinetic environment, they lack the ability to be effective in the information arena. As a result, Turkish military innovation can safely be said to be focused chiefly on kinetic capabilities.

Finally, the societal reflection of military innovation needs to be emphasized. The existence of small, professional units equipped with advanced weapons systems is a development that society widely accepts with enthusiasm. Because the people no longer see their children in at-risk areas within the scope of compulsory military service, of course they welcome this development and have raised their expectations. This is a result of conscripts no longer needing to operate in at-risk areas in the now semi-professional Turkish military. However, every item of news of a martyr is questioned even more so than before because everyone knows about the new military capabilities and expects them to have a casualty-reducing effect. Therefore, military innovation has also had some unintended consequences, and managing these will become much more challenging as soon as society's threat perception decreases.

These explanations reveal threat assessment and operational requirements to be the leading factors in Turkish military innovation. However, these two driving factors were not enough to initiate military innovation, and thus the changes remained as small-scale reform and modernization movements for many years. The enabler that facilitated military innovation was the achievement of the domestic production of high-tech weapons systems. These systems also changed military organization and doctrine, as they have also resulted in increased military effectiveness. In short, every innovation has been seen as a savior for eliminating threats in a country that has longed for a solution to its security problems. But this explanation is not enough to fully understand Turkish military innovation. These innovations became possible thanks to certain novel methods and latent factors that have been successfully applied for the first time in Türkiye.

The Latent Sources of Turkish Military Innovation

How were projects such as UAV and MİLGEM National Warship carried out in a country where domestic politics are often disordered, social polarization is peaked, and even collective action and teamwork are little known? The information Eker and Çam obtained from primary witnesses shows that anger and deprivation turned into motivation in both projects, in which civil-military cooperation as well as solidarity were found between people and bureaucrats from opposite political and ideological poles. The main unifying factor here is both the embargoes as well as the size of the internal threat. In other words, a solid commitment was found in the people who embarked on these projects. The projects' success therefore required dedicated leaders as well as collective work. Ultimately, these groups were able to produce the technology that triggered military innovation. However, the expression of anger and deprivation here may remain an ambiguity to the Western reader. Therefore, the following anecdote from an aircraft engineer is considered to be an excellent example for explaining the mentality of military and civilian experts in Türkiye:

> The most problematic and constantly broken part of the fighter aircraft during the 1950s was the cryogenic system. It was a US-made aircraft, and Turkish engineers were strictly forbidden from opening it. So, we had to send it to the USA for repair, which cost a considerable amount each time. Finally, one day we decided to open it and noticed that the broken part was just a simple electric oscillator that can be found even in household machines. While this was frustrating, it also stimulated us to produce our own systems.[4]

Such experiences are not claimed to trigger the development of defense industry and military innovation; however, one can assume that their cumulative impact resonates with military and civilian decision-makers. Therefore, it cannot be denied that these invisible factors have a role in facilitating military innovation.

In conclusion, the following can be said about the latent sources of Turkish military innovation. First of all, they emerged in an environment independent of the political and ideological fights the country was experiencing. Bureaucrats and technocrats from opposite poles as well as political and military leaders were observed to cooperate behind the scenes. While this may seem like an ordinary phenomenon for any developed country, it is a revolutionary occurrence in a developing country constantly dealing with internal and external crises.

Another feature is the example of civil-military cooperation. Leaving aside that civilians were actually permitted to enter a military barracks, the fact that civilians worked together with the soldiers in the field is a triggering factor in itself. This method has become the norm in the two main projects featured in this book as well as in many other military technology projects. Therefore, in a country known for problematic civil-military relations, the ability of civilians and soldiers to work in harmony for a common purpose is a game-changing factor in military innovation.

Finally, the dedication of the civilians and soldiers who believe in military innovation needs to be considered. While the MİLGEM Project is an example of this in the military field, the UAV development sets an example in the

Table 10.1 The Factors, Enablers, and Objectives in Turkish Military Innovation

Main driving factors	Enablers	Objectives
Threat assessment Operational environment Technology	Investment in the defense industry Investment in human capital Technology Civil-military cooperation Cohesive social networks	Military effectiveness Elimination of embargoes and sanctions

Created by the author.

civilian domain. The presence of dedicated high-level officials and business-people can be seen in both cases. Interestingly, these people have personally witnessed embargoes and restrictions and learned lessons that guided them toward innovation.

To conclude, as summarized in Table 10.1, threats and operational requirements appear as the main driving factors in Turkish military innovation. However, these driving factors alone are not enough. Technology should be noted to have a dual role at this point: It is both innovating and innovated. The factors facilitating this process are seen as human resources, R&D investments, and the civil-military cooperation that the country had not known until recently. The primary purpose appears to be military effectiveness and the elimination of the negative impact from embargoes and restrictions.

It is still too early to claim military innovation as a complete success. The search for a silver bullet can often lead to exaggerated expectations in society as well as in political and military decision-makers. However, considering the open-ended and continuous nature of military innovation, one should not forget that the most important phenomena are the educated civilian and military people who are able to form cohesive social networks, because only in this way can military innovation become productive and sustainable.

Notes

1 For an attempt to formulate the driving factors see Rob Sinterniklaas, *Military Innovation: Cutting the Gordian Knot*, Research Paper:116, Netherlands Defence Academy, Breda, 2018.
2 T. Farrell, T. Terriff, "The Sources of Military Change", in (eds.), T. Farrell, T. Terriff, *The Sources of Military Change: Culture, Politics, Technology*, Lynne Rienner Publishers, London, 2002, pp. 13–14.
3 Bruno Latour, *Science in Action: How to follow scientists and engineers through society*, Harvard University Press, Cambridge, 1987, p. 249–250.
4 Interview with Air Force Captain (Deceased in 2015) Acar Çoşkun, Siteler, Ankara, April 20, 2013.

References

Air Force Captain Acar Çoşkun (Deceased in 2015). Interview by the author, April 20, 2013, Ankara.

Sinterniklaas, Rob, *Military Innovation: Cutting the Gordian Knot*, Research Paper:116, Netherlands Defence Academy, Breda, 2018.

Farrell, Theo and Terriff, Terry "The Sources of Military Change", in (eds.) Theo Farrell and Terry Terriff *The Sources of Military Change: Culture, Politics, Technology*, Lynne Rienner Publishers, London, 2002, pp. 3–20.

Bruno Latour, *Science in Action: How to Follow Scientists and Engineers through Society*, Harvard University Press, Cambridge, 1987.

Index

Page numbers in *italics* refer figures, **bold** refer tables and page followed by n refer notes.